ANIMAL
EXPLORATION
LAB for KIDS

ANIMAL EXPLORATION LAB for KIDS

52 Family-Friendly Activities for Learning About the Amazing Animal Kingdom

MAGGIE REINBOLD, M.S.

QUARRY

Brimming with creative inspiration, how-to projects, and useful information to enrich your everyday life, Quarto Knows is a favorite destination for those pursuing their interests and passions. Visit our site and dig deeper with our books into your area of interest: Quarto Creates, Quarto Cooks, Quarto Homes, Quarto Lives, Quarto Drives, Quarto Explores, Quarto Gifts, or Quarto Kids.

First Published in 2020 by Quarry Books, an imprint of The Quarto Group,
100 Cummings Center, Suite 265-D, Beverly, MA 01915, USA.
T (978) 282-9590 F (978) 283-2742 QuartoKnows.com

Quarry Books titles are also available at discount for retail, wholesale, promotional, and bulk purchase. For details, contact the Special Sales Manager by email at specialsales@quarto.com or by mail at The Quarto Group, Attn: Special Sales Manager, 100 Cummings Center, Suite 265-D, Beverly, MA 01915, USA.

10 9 8 7 6 5 4 3 2 1

ISBN: 978-1-63159-732-9

Digital edition published in 2020
eISBN: 978-1-63159-733-6

Library of Congress Cataloging-in-Publication Data

Reinbold, Maggie, author.
Animal exploration lab for kids : 52 family-friendly activities for
  learning about the amazing animal kingdom / Maggie Reinbold.
ISBN 9781631597329 (trade paperback) | ISBN 9781631597336 (ebook)
1. Animals--Juvenile literature.
LCC QL49 .R384 2020 (print) | LCC QL49 (ebook) | DDC 590--dc23

LCCN 2020000171 (print) | LCCN 2020000172 (ebook) |

Design: Samantha J. Bednarek
Cover Images: Bradford Hollingsworth
Page Layout: Samantha J. Bednarek
Photography: Bradford Hollingsworth, except pages 15 (lower right), 16 (lower right), 23 (lower right), 31 (lower right), 35 (lower right), 37 (lower right), 39 (lower right), 41 (lower right), 45 (lower right), 47 (lower right), 49 (lower right), 50 (lower right), 51, 53 (lower right), 55 (lower right), 57 (all but lower left), 61 (lower right), 63 (lower right), 65 (lower right), 67 (lower right), 69 (lower right), 71 (lower left and lower right), 73 (lower right), 77 (lower right), 79 (lower right), 83 (lower left and lower right), 85 (lower left and lower right), 87 (lower left and lower right), 89 (lower left, lower right, top right), 91 (lower left and lower right), 94 (lower right), 99, 101 (lower right), 103 (lower right), 105 (lower right), 107 (lower left and lower right), 111 (lower left and lower right), 113 (lower left), 114, 117 (lower left, middle, right), 119 (lower right), 125 (lower right), 127 (lower right), 129 (lower right), 131 (lower right), 133 (lower right), 135 (lower right) courtesy of Shutterstock

Printed in China

Dress appropriately when interacting with animals and bugs. Use bug repellents according to label directions. Some animals and bugs are dangerous—get to know the ones in your area that are and do not touch or handle them. If you have a severe allergic reaction such as difficulty breathing (anaphylaxis) or are bitten by a dangerous animal or insect, seek medical attention immediately. The publisher and author assume no responsibility for bites, stings, or any other injuries that may occur while performing these labs.

For Brad and Wren and Phoebe,
my fellow nature explorers
and the loves of my life

CONTENTS

Introduction **8**
Overview **10**

INTRODUCTION

For as long as I can remember, I have held a deep love and respect for animals. My very best memories are of time spent in nature, exploring and observing wildlife. From the vivid memory of touching noses with a curious deer through the screen of our family camper to hovering high above a flock of flamingoes three-million strong in Kenya, animals have provided a constant source of wonder in my life.

Sharing this love of animals with my daughters has been an absolute joy and I have witnessed on multiple occasions the same sense of amazement in them. I remember Wren being completely transfixed as Kenai the wolf at the San Diego Zoo howled into the night sky. The absolute thrill in Phoebe's eyes when she held her first side-blotched lizard. Researching and writing this book provides an opportunity to share my love of animals with many more children than just my own, and for that I'm very grateful.

I believe to my core that understanding other species is the first step in caring about them and ultimately conserving them. My hope is that the activities in this book provide moments of discovery and amazement that then lead to compassion and a strong sense of commitment. Because when you take a closer look at the uniqueness of individual species, you illuminate an important characteristic that binds them all: their vulnerability to the pressures of a rapidly changing world. There has never been a more important time to develop a love for the incredible creatures around us, so let's explore the animal kingdom!

OVERVIEW

This book is full of fun activities designed to enhance your understanding of, and love for, animals. From exploring the techniques that researchers use to *study* animals to discovering the traits and behaviors that make each species unique, you will learn to see your animal neighbors in a new light. You will build on the discoveries of others to think up new ways to support and conserve the amazing creatures around you.

Before you set out on your journey, it's important to start with a pledge to **DO NO HARM**. You will have lots of fun discovering new things about animals and the scientific process, but you must never hurt or harass animals, or deprive them of the things they need. Many of the lab activities have you assume the role of an animal so you can discover and better understand its traits and perspective. That is an important step in the process of appreciating animals and their needs, now and into the future.

It's also important not to put *yourself* in any danger, so observing animals without touching them, for example, is a best practice unless you are certain it's safe to do so. While having fun with the labs in this book, be sure not to put yourself or others in harm's way.

Each lab is designed to help you build new knowledge and skills around animal science and is broken into sections:

→ **Materials** lists all the things you'll need to conduct each lab.

→ **Safety Tips & Helpful Hints** provides additional guidelines and insights for successfully conducting each lab.

→ **Procedure** details the individual steps in each lab so you'll know just what to do.

→ **The Science Behind the Fun** provides a simple description of the science that supports the lab and other background information.

→ **Creative Enrichment** helps you think about how to take your experiment further.

When scientists set out to better understand the world around them, they often start by asking a question based on their observations. This question leads them on a pathway through the scientific process, where they will formulate a hypothesis (basically an educated guess), design an experiment, collect and analyze data, make changes to their hypothesis based on feedback from their peers, and present their findings to the community.

One vital aspect of scientific work is making sure that others can replicate the process and results, since facts are never based on an outcome that is only observed once. For example, if you saw your cat swallow a rubber band one day, you would not then announce to the whole world that the primary diet of all cats is rubber bands. You would be claiming that a single behavior by one (not very smart) cat that was only observed one time by one researcher (you) and not widely observed by others in the scientific community was a scientific fact. That is definitely not how science works!

LAB NOTEBOOK

During this journey to better understand and appreciate animals, you will use a lab notebook (also known as a science journal) to record your process and results clearly for yourself and others. Your lab notebook is the special place where you'll keep all the information and personal notes about your experience with each lab. Using any kind of composition book or spiral notebook, here is the kind of information that you'll want to record for each lab:

→ **Date and Time:** It's always important to record *when* you conduct your experiment, so you'll be able to remember the time of day, date, month, and year. It can also help you think through your findings; how, for instance, doing your experiment during the winter instead of the summer might have affected your results.

→ **My Question(s):** Write down what you're trying to figure out by doing the experiment. What do you want to learn? State your hypothesis: based on what you know, what do you predict will happen?

→ **My Procedure:** Describe what you are going to do to test your hypothesis. This can be a simple list of steps that you plan to follow during the experiment, but the more information the better so other researchers can try it too.

→ **My Results:** Record what happens during the experiment, any observations you make, and the data you collect—numbers, drawings, photos, checkmarks in a column ... it really depends on the experiment.

→ **My Conclusions:** Try to make sense of your results by stating what you think they mean and why you got the results you did. What factors might have affected your results? Did anything unexpected happen?

→ **Future Research:** Record additional questions that you uncover through the experiment and ideas for extending the activities. Oftentimes the answers to your scientific questions end up leading to more questions, but that's the exciting part ... science is never officially *done*!

STUDYING ANIMALS

Zoology is generally defined as the study of animals, and it has been practiced and perfected by scientists over many centuries. From groundbreaking pioneers, such as Charles Darwin, to modern-day heroes, such as Jane Goodall, animal scientists work tirelessly to understand and document the traits, behaviors, distributions, and needs of the millions of animal species on our planet.

How animals are related, what they eat, where they live, when and how many babies they have, how they move, how much they sleep, where they hide, and what they need to thrive are just a few of the things we know thanks to animal scientists. But there is still so much more to find out! Researchers discover new species every year, and as our world continues to change, we will need legions of new scientists to study and preserve the biodiversity that is so critical to our planet's well-being.

The labs in this unit help you build your own research skills by having you experiment directly with some of the tools and techniques that researchers use to study animals. From secretly capturing animals on camera to sifting through clues they leave behind, this unit lets you become an animal researcher and collect important data you can share with your community.

LAB 01

ANIMALS ALL AROUND US

Learn about your neighbors with an animal artifact scavenger hunt.

MATERIALS

→ Handheld magnifying glass
→ Digital camera or handheld device with camera (tablet, smartphone, etc.)
→ Collecting bin (a basket or bucket will work)
→ Garden or work gloves (optional)

SAFETY TIPS & HELPFUL HINTS

→ For this first activity, it's important to keep an open mind. If you have a set idea of what you're looking for, you're likely to miss interesting clues about the animals living around you. For this activity, focus on looking for artifacts that animals leave behind rather than on any live animals you may see. Common findings include tracks (prints), scat (poop), leftover food, bones, feathers, fur, claws, webs, burrows, shed skin, and nests.

→ Be cautious about handling the artifacts that you find; for example, use gloves when examining scat or leftover animal food to avoid getting germs on your hands.

PROCEDURE

1. Give yourself at least 30 minutes to explore the areas around your home and neighborhood. Work slowly and observe the habitat—rushing doesn't work well for this activity! Be creative about *where* you look: peek inside bushes, turn over rocks and logs, and climb up a tree (carefully!) to see what you find (**Fig. 1**).

2. Whenever you come across an animal artifact, sketch it in your lab notebook, recording all the details you can (size, shape, color, texture, smell, etc.). Of course, never taste the artifact (**Fig. 2**)!

3. If the artifact is no longer being used by any animals, then go ahead and put it in your collecting bin. If it might still be in use, such as a web or nest, then take a picture of it with your camera (you can print your artifact photos later and include them with your other findings) (**Fig. 3**).

4. Record what kind of animal you think left each item behind (insect, bird, mammal, etc.).

Fig. 1

Fig. 2

Fig. 3

THE SCIENCE BEHIND THE FUN

→ Animal researchers rely heavily on their ability to make good observations and on their awareness of what is around them. They use many creative techniques to study animals—some that you'll explore later in this unit—but they also make important discoveries by simply going outdoors to explore the habitat. While there, scientists who pay attention to their surroundings can learn incredible things about animals. They can find out which animals live in which habitats, they can discover that some animals actually use tools, and they can even document brand new species of animals! In this lab, you'll hone your observation skills and begin to consider the many fascinating traits that link animals together.

CREATIVE ENRICHMENT

→ Researchers spend a lot of time discussing their findings with others. To share your observations, consider arranging your scavenger hunt discoveries into an animal artifact display. You might choose to arrange your artifacts by animal group or have all similar items in one area. There is no right or wrong way. Just have fun and be creative!

LAB
02

CLASSIFYING ANIMALS

Use shared traits to group animals like scientists do.

MATERIALS

→ Computer with internet connection and browser

SAFETY TIPS & HELPFUL HINTS

→ Before entering any search terms in Google, ask a parent to help you enable the "SafeSearch" feature, which blocks any inappropriate images.

PROCEDURE

1. In your lab notebook, draw seven columns. Label them Mammals, Birds, Reptiles, Amphibians, Fishes, Arthropods, and Other Invertebrates. Under each title, list a few traits unique to that group **(Fig. 1)**.

2. Search Google images with the term "animal." Click through the resulting photos and record the name of the pictured animal in its corresponding column. For example, under which animal group would you list a tiger? A clownfish? A rattlesnake? A bald eagle? A butterfly? A black widow spider? Categorize the first 50 animals, reminding yourself of the traits used to make each classification. Which animal group appeared most? Were any groups missing **(Fig. 2)**?

3. Now use the search term "arthropod." In the Arthropods column of your notebook, record the first 20 animals that appear.

4. Lastly, use the search term "invertebrate." Categorize the first 20 animals that appear in the corresponding columns of your table (Arthropods or Other Invertebrates) **(Fig. 3)**.

CREATIVE ENRICHMENT

→ Now that you know the traits of an arthropod (exoskeleton, segmented body, pairs of jointed legs), consider learning more about other groups of invertebrates. What are mollusks? Echinoderms? Poriferans? Many great resources can help you learn about invertebrate diversity, including the Encyclopedia of Life (eol.org). And remember to be grateful for these little creatures that make life possible!

Fig. 1

Fig. 2

Fig. 3

THE
SCIENCE
BEHIND THE FUN

→ To better understand animals and how they are related, scientists use shared traits to group and name them. If, for example, an animal has a backbone, hair or fur, produces milk for its young, reproduces through live birth (instead of laying eggs), and is warm-blooded, then it is classified as a mammal. Animal classification starts at the highest level, KINGDOM, and goes all the way down to the most specific level, SPECIES. Within the animal kingdom, species are grouped into two high-level categories based on whether or not they have a backbone: vertebrates (yes) and invertebrates (no).

In this lab, you will classify animals within seven high-level groupings: Mammals, Birds, Reptiles, Amphibians, Fishes, Arthropods, and Other Invertebrates. First, consider vertebrates. You have already reviewed the classifying traits of mammals (hair or fur, produces milk, live birth, warm-blooded). If an animal has feathers and wings, lays eggs, and is warm-blooded, then it is classified as a bird. If an animal has scaly skin, lays eggs, and is cold-blooded, then it is classified as a reptile. If an animal has moist skin, goes through metamorphosis, and is cold-blooded, then it is classified as an amphibian. If an animal breathes through gills, lives in water, lays eggs, and is cold-blooded, then it is classified as a fish. There are, of course, exceptions to many of these generalized rules. For example, some snakes and some fish actually give birth to live young instead of laying eggs.

Now consider invertebrates. There are *vastly* more invertebrate animals on this planet than there are vertebrates. In fact, more than 97 percent of all animal species fall into this category. Invertebrates tend not to be the first animals that come to people's minds, as you'll see from the activities in this lab, but human survival on this planet would be impossible without them. For our purposes, you'll focus mostly on Arthropods, the largest invertebrate group, containing animals that have an exoskeleton, a segmented body, and multiple pairs of jointed legs (butterflies, beetles, spiders, centipedes, scorpions, crabs, and more). But do keep in mind that there are many other interesting invertebrates to discover (snails, sea stars, octopuses, worms, corals, clams, jellyfish, and more). In this lab, you'll use shared traits to classify animals like scientists do.

PRODUCTIVE PITFALLS

Discover local arthropods with simple cup traps.

MATERIALS

→ Large plastic cups (with holes poked in the bottom of each to drain rainwater)
→ Hand shovel
→ Small shade structure (four popsicles sticks and a cloth napkin held together with rubber bands)
→ Handheld magnifying glass
→ Ruler (with centimeters)
→ Clipboard

SAFETY TIPS & HELPFUL HINTS

→ When deploying pitfall traps you *must* check them often enough so no visiting animal is caught for too long. The minimum is once per day, but more often is better.

→ Choose your location wisely so your pitfall traps don't become a tripping hazard.

→ Remember to be very gentle when handling your visiting animals. Release each organism near where it was caught.

PROCEDURE

1. Spend some time in your yard or neighborhood before you decide where to place your pitfall traps. Areas with plants, moist soils, and at least some shade are likely to attract the most visitors.

2. Use a hand shovel to dig a small hole for each plastic cup (pitfall trap) that you want to deploy in your chosen location **(Fig. 1)**.

3. Make sure that your hole is a good fit for the size of cup that you're using and that the surrounding dirt comes right up to the rim of the cup once it's buried in the ground. The holes that you poked in the cup will help rain leak out so the arthropods you collect don't drown **(Fig. 2)**.

4. Make a square on the ground with four popsicle sticks and drape them with a cloth napkin held in place with a rubber band at each corner. This simple structure will protect visiting animals from direct sunlight **(Fig. 3)**.

5. When you find a visitor during your frequent trap checks, remove the cup from the ground and either observe the animals in the cup or gently pour them onto your clipboard. Record your observations of the animals' physical features and behaviors in your lab notebook and sketch each animal, including labels for each body part. Measure the length and width of your visitors so you can calculate average size and the size range for each species.

6. Release the animals near where you caught them.

Fig. 1

Fig. 2

Fig. 3

THE
SCIENCE
BEHIND THE FUN

→ Pitfall traps, also called cup traps, are a simple way to survey for small animals living in an area. When deployed responsibly, they let you capture and observe animals safely, making sure that each creature can be released unharmed. Scientists using this method document the diversity and abundance of species throughout the year and across habitat types, especially for small, ground-dwelling creatures, such as arthropods, reptiles, and amphibians. When an animal is caught in the trap, researchers gently collect a variety of data before putting the creature back into its habitat. For example, scientists studying lizards record the weight and length of each animal, taking photographs to document markings and coloration.

Researchers must obtain specialized permits for pitfall traps in the wild, and must also assess the ecosystem before beginning their research. To ensure the safety and comfort of captured animals, a lot of thought goes into trap installation and deployment. For example, researchers must keep animals sheltered from weather conditions and provide them with water and hiding spots or materials for protection and warmth. In this lab, you'll use simple cup traps to document local arthropods.

CREATIVE ENRICHMENT

→ Once you are familiar with the use and upkeep of pitfall traps, try installing them in different location types so you can compare the number and kinds of organisms that visit each place. For example, do you find different visiting animals in areas with native plant species (those that naturally occur in your region) as opposed to ornamental or introduced plant species? How about areas with sandy soil as opposed to loamy or clay soils? You'll have endless options for comparison!

CAUGHT ON CAMERA

Use a hidden trail camera to find out who lives nearby.

MATERIALS

→ Motion-activated trail camera
→ Batteries
→ Memory card
→ Mounting stake, tree, or other sturdy structure

SAFETY TIPS & HELPFUL HINTS

→ Trail cameras, or cams, are easy to purchase online at a relatively low cost.

→ If you don't have a yard or nature space around your home, ask someone who does if you can deploy your trail cam on their property for a few days to document *their* visiting species.

→ Privacy is very important, so be sure to avoid placing your trail cam in areas where a lot of people hang out or on property you don't have permission to use. No one likes being photographed without their knowledge, so be sure to tell your family members about the trail cam before it's deployed.

PROCEDURE

1. Get your trail cam ready to use—add batteries, a memory card, and choose your preferred settings as directed in the instructions that came with it.

2. Spend a few days paying extra attention to what animals use your yard as habitat and where they like to hang out. Use your observations to decide where to place your trail cam—high-traffic areas will trigger more wildlife images.

3. Fasten your trail cam to a wooden stake in the ground or to another sturdy structure, such as a tree or fence, in your chosen location **(Fig. 1)**.

4. If you have lots of animal activity, you will get lots of images, so check the camera every day. Most trail cam models let you review the images right on the internal screen, but you'll also want to download the images so you can review them and print your favorites **(Fig. 2)**.

5. You can change the location of your camera every few days to maximize your chances of documenting *all* the animals using your yard. Most trail cams have a nighttime function, so you can also document nocturnal animals around your home **(Fig. 3)**!

Fig. 1

Fig. 2

Fig. 3

THE SCIENCE BEHIND THE FUN

→ Trail cams are specialized cameras with sensor technology that researchers deploy at field sites around the world in order to collect data through photographs of wildlife and associated ecosystems. They allow researchers to study with little to no impact on wildlife, and they can also be deployed at remote sites for extended periods of time without the researchers being present. The cameras can be programmed to capture images when triggered by motion or at set intervals throughout the day, depending on the given research questions. This functionality allows scientists to address many questions about wildlife, from documenting diversity and abundance of species in poorly studied regions to how habitats recover following natural disasters, such as wildfires and floods. In this lab, you'll use a hidden trail camera to see who lives near you.

CREATIVE ENRICHMENT

→ Sharing your findings is an important part of the scientific process, so consider turning your results into easy-to-interpret, colorful graphs that you can share. How many individual animal images did you capture? How many photos did you capture of different animal groups (reptiles, birds, mammals, etc.)? There will be a lot to share!

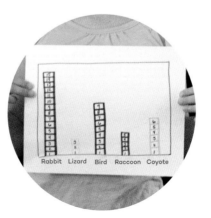

CATCHING TRACKS
Build a track station to see who's passing through.

Fig. 3

MATERIALS
→ Fine-grained sand
→ Mineral oil
→ Handheld magnifying glass
→ Ruler

SAFETY TIPS & HELPFUL HINTS

→ Select a spot for your track station where people aren't likely to pass through so you won't constantly be recording shoeprints!

→ Check the forecast before you set your grid so you know if the weather will cooperate. Even a little rain can ruin your data collection.

→ You may want to make the area for your track station flat and flush with the surrounding ground, as long as the change doesn't deter visiting animals.

PROCEDURE

1. Pitfall traps helped you examine arthropods and your trail cam helped you document species that are large enough to trigger the camera. Now, use this information to decide where to set up your track station. You'll be studying animals that move on the ground and are heavy enough to leave prints, so choose a location that you think experiences a lot of animal foot traffic. The spot must be flat and even, so avoid things such as grass and gravel.

2. Blend your fine-grained sand with mineral oil in a bucket or mixing bowl, adding oil until the sand is shapeable but not too wet or saturated **(Fig. 1)**.

3. Spread the oiled sand into a rectangular grid in your location of choice. The grid doesn't need to be huge, but the larger it is, the higher the probability that something will walk through it. Consider using a grid that is about 2 feet wide and 3 feet long (0.6 m by 0.9 m). The sand should be at least a ½ inch (1.3 cm) deep in order to capture tracks adequately **(Fig. 2)**. Leave the track station untouched for a whole day (24 hours).

4. Once you have tracks in the sand, carefully record their length and width and sketch them in your field notebook **(Fig. 3)**. Many guidebooks and online resources can help you identify the tracks, especially if you're just keying them to animal group (mammal, bird, reptile, etc.) rather than particular species.

Fig. 1

Fig. 2

→ Researchers use track stations to document the presence and activity of animals in a given area. The track station is covered with a substance, such as chalk, sand, or ash, that allows animals to leave behind prominent footprints. Researchers can identify species and sometimes even gender and age through tracks. To attract animals to the station, researchers also sometimes use bait, such as food or scents. In some interesting cases, scientists have successfully attracted large mammals, such as leopards and jaguars, with popular perfumes! In this lab, you'll deploy a track station to see who's passing through your area.

CREATIVE ENRICHMENT

→ Researchers sometimes use bait, such as food or scents, when studying animals through a track station. After re-smoothing your track station sand, try putting out different types of bait. Do you get more animals? Different animals? Some good things to try are cat food, canned tuna fish, dog kibble, and fish flakes. Or, you can try perfume or other household scents. If you're lucky enough to get a deep, well-defined track at your station, consider using plaster of Paris to create a keepsake track (follow the directions on the product) from your study that you can even paint and decorate!

LISTENING IN

Use a digital audio recorder to tap into the secret conversations of animals.

MATERIALS

→ Digital audio recording device
→ Mounting stake, tree, or other sturdy structure

SAFETY TIPS & HELPFUL HINTS

→ You can purchase a variety of affordable digital audio recording devices online. A smartphone or MP3 player will also work, but they will be expensive to replace if they get lost or broken.

→ Check the forecast and humidity levels (the lower the better) before deploying your recording device, since they can be quite sensitive to wet weather conditions.

→ As noted in Lab 4, privacy is very important, so do not place your audio recorder in areas where people tend to hang out (we're trying to record sounds from animals that *aren't* humans!). No one likes to be spied on, so be sure to tell your family members about the recorder before it's deployed.

PROCEDURE

1. Most digital audio recorders are ready to use right out of the box, but read through the instructions before you turn it on and start recording. Ideally, make short recordings at different times throughout the day and at night. The more you record, the more data you'll have—and have to review (**Fig. 1**)!

2. Since the only animals you'll capture with this kind of technology are those that communicate through sound, you should be able to use this knowledge to place your audio recorder. It doesn't have to be on the ground like your tracks station, so perhaps consider an elevated spot— a tree?—where chatty species tend to spend time.

3. Secure your audio recorder to a sturdy structure in your chosen spot, and protect it as necessary from getting wet (**Fig. 2**).

4. After a few sessions, download your captured sound files and get ready to play detective. It will take time to review all the data, so be patient and take breaks when you need them (**Fig. 3**).

5. Each time you discover a new animal sound, consider what type of animal might have produced it. Record your ideas in your lab notebook. Many online resources and apps can help you identify the calls so you don't have to figure it out by yourself.

Fig. 1

Fig. 2

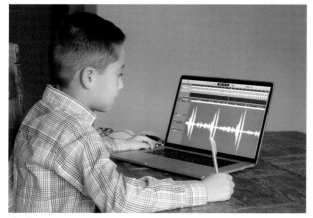

Fig. 3

CREATIVE ENRICHMENT

→ Consider editing your animal sounds into one continuous "song" that you can share. Many apps can help you do this, and certainly don't hesitate to ask for help from a parent or tech-savvy friend.

THE
SCIENCE
BEHIND THE FUN

→ Audio recorders, also called bioacoustic recorders, are small microphones set to pick up sounds of animals in a particular region or habitat of interest, letting researchers develop a greater understanding of species diversity, population size, and social interactions between animals. Researchers don't have to be there in person for them to work, and the recorders can be programmed to gather information over long or short intervals. They are indispensable in areas where visual surveys are difficult, such as in the thick vegetation of tropical forests. Audio recorders can capture audio animal communication that falls outside of the normal human hearing range, such as low-frequency infrasound, and scientists also use them to monitor the impacts of human-made noises on surrounding natural areas. They have even captured recordings of species previously thought to be extinct! In this lab, you'll use a digital audio recorder to capture animal conversations.

BEHAVIOR INVENTORY

Record animal behavior like a scientist!

MATERIALS

→ Timer

SAFETY TIPS & HELPFUL HINTS

→ You can study the behavior of animals in your yard, house (pets count!), a local park, or even at a nearby zoo or aquarium. If all else fails, many zoos maintain animal cams that can be accessed online, so you can record your observations on the computer screen.

→ As you set out to observe and record animal behavior, it is important to remember that, as the researcher, you are not trying to influence the animal's behavior. In addition to the principle of **DO NO HARM**, you also don't want to sway your scientific results. Do your best to stay quiet and still as you make your observations.

→ Choose a comfortable spot—you might be there for a while! Consider sitting or standing in the shade and not in the path of others, so you can concentrate on the task at hand.

PROCEDURE

1. Before you start your timer, choose one animal and watch it for at least 5 minutes, listing its behaviors in your lab notebook. Is it sleeping, resting, eating, drinking, playing, smelling, or just moving? Describe each behavior briefly so others can understand what you're observing. You have just built an ethogram **(Fig. 1)**!

2. Once you have a pretty good idea about the range of behaviors your animal exhibits, you're ready to collect some data. Make two columns in your lab notebook for recording the animal's behavior at set time intervals (0 seconds, 30 seconds, 60 seconds, 90 seconds, etc.). For beginners, it's smart to choose intervals of at least 30 seconds so you have time between each recording and you don't get overwhelmed **(Fig. 2)**.

3. Set your timer for 30 seconds and press start. Write down what behavior the animal is doing at that very moment (at 0 seconds). After 30 seconds, record what behavior it's then doing. Restart your timer for another 30 seconds and continue this until you have completed 3 minutes of observation. You should have seven behavioral observations in your table **(Fig. 3)**.

4. Let a couple minutes go by and then start a new 3-minute trial. Running more trials will make your conclusions about the animal's behavior more accurate.

5. Now it's time to analyze your data. What is the most common behavior that you observed? How might your behavioral data change at a different time of day? Did anything surprise you about the animal's behavior?

Fig. 1

Fig. 2

Fig. 3

THE SCIENCE BEHIND THE FUN

→ The study of animal behavior is called *ethology*, and it provides researchers with valuable information for the management and conservation of animals. One important tool for studying behavior is an ethogram. To build an ethogram, scientists list and describe the various behaviors of an animal so they can be studied and understood. Ethograms help researchers collect and compare data about individual animals, groups of animals, and even across species. Once a comprehensive ethogram is built, data about the needs, preferences, and uniqueness of animals can be collected using a variety of sampling methods so researchers can make informed decisions about care and conservation. In this lab, you'll study animal behavior by building an ethogram.

CREATIVE ENRICHMENT

→ Consider whether the behavior observed is typical of that species. For example, if you made observations about your dog, would you expect the exact same behavior in your neighbor's dog? Try observing more than one individual within the same species to see if you get similar results. Perhaps turn this interesting data into colorful graphs or charts.

INCREDIBLE ANIMAL ADAPTATIONS

One important driving force throughout the animal kingdom is the need to live long enough to pass genes on to the next generation. Species have evolved an incredible array of interesting traits to help accomplish this. Some are behavioral (an opossum playing dead or a squirrel storing acorns for winter) and some are physical (the striped fur of a tiger or the horn of a rhino). Useful traits can help animals capture prey, attract a mate, move efficiently, escape from predators, or stay comfortable in their habitat.

A trait that helps an animal survive better in its natural environment is called an adaptation. Adaptations can arise in a variety of ways, and those that help animals survive long enough to reproduce can become widespread in populations over time—it may eventually become a species' defining characteristic.

The labs in this unit will help you discover how physical traits can lend advantages to animals as they work to survive and thrive. From investigating the role of beak shape in feeding ecology to experimenting with the insulating power of blubber, this unit gives you a firsthand look at body features that help animals succeed in a variety of niches.

LAB 08

OUR AMAZING OPPOSABLE THUMB

Try to perform basic tasks without this important primate adaptation.

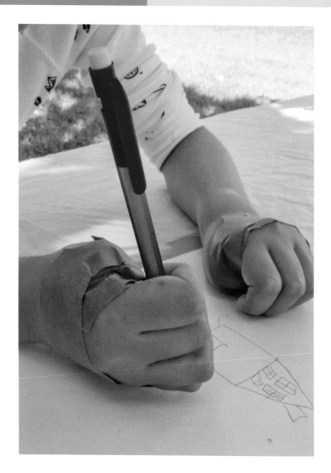

MATERIALS

→ Masking tape or painter's tape
→ Timer
→ Shoes with shoelaces
→ Toothbrush and toothpaste
→ Bowl of ice cream and spoon
→ Hairbrush
→ Pencil and paper
→ Help from a friend or family member

SAFETY TIPS & HELPFUL HINTS

→ For this lab, be sure not to perform any tasks—such as riding a bike or driving a go-cart—that can be dangerous to do without thumbs!

PROCEDURE

1. Have a friend or family member tape both of your thumbs down in a resting position across the palm of your hand **(Fig. 1)**.

2. Using the timer, have a friend or family member record how long it takes you to do a variety of simple tasks. Try tying your shoes, brushing your teeth, eating ice cream with a spoon, brushing your hair, writing your name, picking coins up off the floor, and turning a doorknob **(Fig. 2)**.

3. Remove the tape, record your times for each task in your lab notebook, and then time yourself doing each task again, with your amazing opposable thumbs back in action **(Fig. 3)**.

Fig. 1

Fig. 2

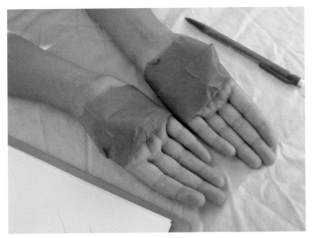

Fig. 3

THE SCIENCE BEHIND THE FUN

→ To explore how certain traits can lend advantages to animals as they work to survive and thrive, let's first examine an adaptation that humans possess: our amazing opposable thumb. Being able to use our thumb to touch the other fingers on our hand makes almost every aspect of daily life much easier. Opposable thumbs are quite rare in the animal kingdom, present in nearly all primates (including humans) and only a few other species. Other animals use various body structures to help them grasp objects, but your amazing opposable thumb sets the standard. In this lab, you'll try to perform basic tasks without this adaptation.

CREATIVE ENRICHMENT

→ Believe it or not, some animals (and people, too) are just as good at using their feet to perform basic tasks as they are with their hands! This is especially true for animals that have no hands, such as birds. They can crack open nuts, capture prey items, and build complex nests using only their beaks and the digits on their feet. Try performing some of the basic tasks in this lab using only your feet. Can you write your name? Can you tie your shoes? Not so easy, is it!?

LAB 09

A BEAK FOR EVERY MEAL

Discover how beak shape helps birds thrive in different environments.

MATERIALS

→ Simulated beaks (tweezers, kitchen tongs, clothespin, pipette or syringe)
→ Simulated foods (mini marshmallows [plump grubs], Swedish fish [fish], uncooked rice grains [ants], popcorn kernels [seeds], toothpicks [stick insects], pieces of rubber bands [worms], bowl of sugar dissolved in water [nectar])
→ Simulated stomach (small, non-breakable cup for each player)
→ Large, rimmed baking sheet
→ Timer
→ Players (four works well)

SAFETY TIPS & HELPFUL HINTS

→ Run the simulation using the sugar water last because it tends to be messier. Measure the water level in each cup after this round.

PROCEDURE

1. Before starting the game, draw a data table with seven columns and four rows in your lab notebook. Label the columns with each of the seven simulated foods **(Fig. 1)**. Label the rows with each of the four simulated beak types **(Fig. 2)**.

2. Give each player a simulated stomach (empty cup) and a "beak." Explain that you'll be competing for different food types, using only your "beaks" to pick up food items and place them into your cup. Each feeding round will last 20 seconds, so set your timer.

3. Spread the first food item onto the baking sheet. Say "GO!" as you start your timer. After 20 seconds are up, record the amount of food each beak was able to collect.

4. Repeat the game with each food type until you have collected data for all of them **(Fig. 3)**.

5. Determine which beak worked best for eating each food type. Was one beak type good at capturing ALL the different food items? Consider creating colorful graphs and charts to display the game results **(Fig. 4)**.

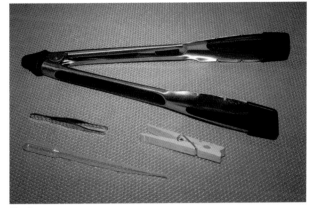

Fig. 1

Fig. 2

Fig. 3

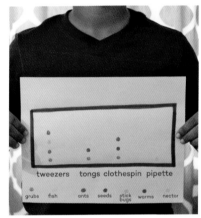

Fig. 4

tweezers   tongs clothespin pipette

grubs fish   ants seeds stick   worms nectar
                            bugs

THE SCIENCE BEHIND THE FUN

→ Often in the animal kingdom, individual physical traits vary greatly across species. In this lab, you'll experiment with simulated bird beaks and a variety of food. Differences in bird beaks are great examples of adaptive variation. In fact, that is what piqued Charles Darwin's interest when he visited the Galapagos Islands in 1835. He was fascinated to find that each island had its own type of finch with different beaks to help take advantage of the food resources available on its own island.

As you think about which beak type works best for each food item, consider which beak worked best at capturing a variety of food. Scientists use the terms "specialist" and "generalist" to describe this phenomenon. A specialist species tends to do well only in a very specific environment and often with a very specific diet. A generalist species is able to do well in a variety of environments and use many different resources. In this lab, you'll discover how beak shape helps birds thrive in different environments.

CREATIVE ENRICHMENT

→ Now that you have experimented using pre-existing "beaks," try designing your very own beak! Think about which design elements proved most helpful for gathering up the food items. For example, if one beak type seemed too slippery, leading food items to continually fall from its grasp, how would you make a stickier beak? Be creative and have fun!

MULTITASKING TRUNK

See how well your "trunk" works for siphoning water.

PROCEDURE

1. Pour 1 gallon (3.8 L) of drinking water into the first bowl.

2. Connect your two large straws together to make a "trunk" that is more comparable to the size and ratio of a real elephant trunk. To do this, carefully feed the end of one straw into the opening of the other to get a snug fit **(Fig. 1)**.

3. Use your "trunk" to move all the water from one bowl to the other, recording the number of moves it takes to complete the task **(Fig. 2)**. Make sure the water stays in the straw and doesn't reach your mouth. And don't drink the water either! How hard was this task? Did it get harder over time? What muscles were you using **(Fig. 3)**?

MATERIALS

→ Two large bowls
→ Measuring cup
→ Drinking water
→ Two large straws

SAFETY TIPS & HELPFUL HINTS

→ While you could also time yourself for this activity, this might lead to coughing, choking, or water coming out your nose (not fun!). So just take it slow and measure how many individual *moves* it takes to transfer all the water rather than how much *time* it takes.

Fig. 1

Fig. 2

Fig. 3

THE
SCIENCE
BEHIND THE FUN

→ One of the most incredible and versatile adaptations in the animal world, an elephant's trunk allows its owner to feel, lift, touch, smell, feed, throw, siphon, snorkel, grab, spray, and even speak. The elephant's upper lip and nose are combined to form the trunk, complete with two long nostrils running from the face to the tip. It consists of thousands of muscles and allows its owner to detect scents from many miles away. On its own, an average trunk can weigh more than 300 pounds (136 kg) and can have the strength to lift more than 550 pounds (249 kg)! While many people think that elephants can drink through their trunk, this is not the case. But they *do* use their trunk to move water from a source into their mouth. In this lab, you'll discover how easy (or hard) it is to move the water that an average elephant can hold in its trunk (about 1 gallon [3.8 L]) from one place to another.

CREATIVE ENRICHMENT

→ Now that you have recorded how many moves it takes to transfer the water carried by an *average* elephant, try recording the number of moves it takes to transfer the water carried by a large bull elephant! That means you'll now be transferring 2.5 gallons (9.5 L) of water! In addition to recording how many individual moves it takes, also write down any other thoughts that occur to you; did you get tired, did it become difficult, etc.

CAPTIVATING CAMOUFLAGE

Explore the effectiveness of camouflage for keeping animals hidden.

MATERIAL

→ Classic Skittles (minimum 100 of each color, separated into bags or containers) (Fig. 1)

→ Five small bags of classic M&Ms

→ Five white paper plates colored brown, green, orange, red, and yellow (Fig. 2)

→ Small plastic or paper cup for each player

→ Timer

→ Large collecting bowl

→ Players (four works well)

Fig. 1

Fig. 2

SAFETY TIPS & HELPFUL HINTS

→ Be sure everyone washes their hands before starting the game so you can all enjoy the leftover candy!

PROCEDURE

1. Add one bag of M&Ms to each color-labeled container of Skittles.

2. Draw a data table with five columns and four rows in your lab notebook. Label the columns with the five plate colors and label the rows with each player's name.

3. Give each player a cup and explain that you'll all be competing for M&Ms. Each feeding round will last 20 seconds, and the point of the game is to pick up as many M&Ms as you can without getting any Skittles.

4. Set your timer for 20 seconds then pour the red-labeled Skittles container onto the red-colored plate (**Fig. 3**). Say "GO!" as you start your timer. After the 20 seconds are up, record the number *and* color of M&Ms each player collected, along with any red Skittles they grabbed.

5. Reset the timer. Empty each player's cup into the bowl and clean up the plate you used. Pour the next Skittles container onto the corresponding colored plate, repeating the game until you have data for all five Skittles colors.

6. In reviewing your data, look for patterns. For example, during the red round, did players collect more or fewer red M&Ms? Did that pattern hold for all color rounds? Why do you think this happened? What about the blue M&Ms? Did players collect them a lot or not very often? Why do you think this happened?

Fig. 3

CREATIVE ENRICHMENT

→ Now that you have explored how blending in can help animals go unnoticed by potential predators, try using actual animal images! Gather 20 animal pictures from online, magazines, or your own artwork. Place all 20 pictures around a room, blending half of them in with similar colors or patterns (on books, blankets, stuffed animals, lampshades, etc.) and leaving the other half where they do not blend in at all (in the middle of a wall, on the surface of your desk, in the middle of the floor, etc.). Bring someone else into the room and ask them to find all 20 pictures. Which did they discover most quickly? Did hiding the animal pictures amongst similar colors and patterns work to conceal them?

THE SCIENCE BEHIND THE FUN

→ Many creatures in the animal kingdom rely on captivating body coloration to survive. Some use bright and conspicuous color patterns to warn potential predators that they are poisonous or distasteful; scientists call this aposematic coloration. Other animals mimic the aposematic coloration of another species in order to avoid being eaten by predators; scientists call this Batesian mimicry. But by far the most common form of protection through coloration is camouflage, in which animals use colors, patterns, or body position to make themselves harder to see. From butterflies and katydids to seahorses and even zebras, camouflage is an effective strategy.

But not all camouflage serves to keep animals safe from predators. Many use it to hide in order to ambush their prey. Consider the many species of mantises. Some look like dead leaves while others look like flowers, but all are predatory, and they use these incredible colors and patterns to lie in wait.

In this lab, players discover the importance of camouflage for keeping prey (M&Ms) hidden from predators. Your M&Ms will hide amongst Skittles on a colored background.

COOL COLORS

Explore how pale coloration helps many desert animals survive the heat.

MATERIALS

→ Computer with internet connection and browser
→ Black cloth
→ White cloth
→ Digital outdoor thermometer
→ Timer

SAFETY TIPS & HELPFUL HINTS

→ Before entering any search terms in Google, enable the "SafeSearch" feature to block any inappropriate content.

→ You'll need to conduct this lab on a sunny day!

→ It can be dangerous to let yourself get too hot, so use an outdoor digital thermometer for this lab, instead of your own body.

PROCEDURE

1. Do a Google images search for "desert animal" and click through the resulting photos **(Fig. 1)**. What do you notice about the coloration of most desert animals? Why do you think they are like this?

2. Go outside and record a starting temperature from your thermometer in your lab notebook.

3. Place the thermometer in direct sunlight and cover it with the white cloth (find yourself a comfortable spot in the shade). Set your timer for 30 seconds and hit start. When the 30 seconds are up, record the temperature on the thermometer. Continue this for 10 minutes **(Fig. 2)**.

4. Once the 10 minutes are up, bring the thermometer inside and let it cool to the same starting temperature you recorded.

5. Head back outside, and repeat the experiment using the black cloth **(Fig. 3)**.

6. After the second 10 minutes, go have a cool drink while you examine your data.

Fig. 1

Fig. 2

Fig. 3

THE SCIENCE BEHIND THE FUN

→ Life in the desert is hard. Extreme temperatures and lack of water and shade or shelter mean that only well-adapted animals and plants survive there. Consequently, animals in arid deserts have evolved traits and behaviors that help them take the heat.

Dark colors, such as black, **absorb** light energy from the sun and convert it into heat. Lighter colors, such as white, **reflect** light energy from the sun. Darker objects heat up in the sun much more quickly than lighter objects. In this lab, you'll consider the physical coloration of most desert animals.

CREATIVE ENRICHMENT

→ Now that you have compared the absorption and reflection of light energy between black and white, consider testing the properties of other colors. After all, few animals are entirely black or entirely white; desert animals especially are more commonly shades of tan or gray. So, for instance, does pink heat faster than yellow? Be sure to record your data so you can share your findings.

LAB
13

THE GREAT BLUBBER CHALLENGE

Explore the importance of blubber for keeping polar animals warm.

SAFETY TIPS & HELPFUL HINTS

→ Wearing gloves will protect you from the insulating materials, some of which can be quite messy and hard to remove. If you have allergies or sensitivities, choose an appropriate glove material.

→ Refresh the ice water as necessary to keep it cold!

→ Consider doing this lab outside, since it can get messy.

PROCEDURE

1. List each insulation type in your lab notebook, leaving space to record a time.

2. Label a bag for each insulation material, then fill each bag accordingly **(Fig. 1)**.

3. Put on a glove.

4. Put your gloved hand into the first insulation bag, start your timer, and then dunk your insulated hand into the ice water **(Fig. 2)**.

5. Record how long you can stand to keep your insulated hand in the water before it gets too cold.

MATERIALS

→ Bowl or bucket of ice water large enough to dunk your entire hand up to the wrist
→ Plastic bags (snack-sized zipper bags work well for most kid-sized hands)
→ Permanent marker

→ Variety of insulating materials (vegetable shortening, cotton balls, butter, petroleum jelly, foam packing peanuts, etc.)
→ Disposable gloves (latex, nitrile, or plastic food service gloves)
→ Timer

Fig. 1

Fig. 2

6. Repeat this procedure with each insulation type and without any insulation at all, letting your hand return to normal body temperature between trials.

CREATIVE ENRICHMENT

→ Consider doing the lab again using a digital meat thermometer inside each insulation bag, instead of your hand. This method is less personal (you don't get to feel and describe the experience literally firsthand), but it is actually more precise to record digital temperature readings for each insulation material (and without any insulation). How quickly does the temperature on the thermometer drop with each material? In your opinion, which insulation would be most effective at keeping an animal warm while swimming in icy polar seas? Be sure to give the thermometer time to return to air temperature between each trial so that your readings all begin from the same starting point. Consider arranging your results into a simple table or colorful graph to share.

THE SCIENCE BEHIND THE FUN

→ Many animals live in very cold environments, such as near the north and south poles or deep in the ocean. To survive and thrive in the cold, many rely on thick layers of fat directly under the skin, an adaptation called blubber. Blubber not only keeps these animals warm, it provides them with energy and helps them stay afloat. Most people think of blubber as keeping the cold out, but it really acts as an insulator: a material that stops or slows the transfer of heat from one area to another. The fat in blubber does not transfer heat as well as other tissue types, so the heat generated by the animal's body stays trapped. In this lab, you'll explore and compare the effectiveness of different insulation types.

INTRIGUING ANIMAL
BEHAVIORS

Behavior is the way an animal interacts with its environment and with the creatures living around it. Animals exhibit all sorts of intriguing behaviors, some they know from birth and some they learn through experience. Some behaviors are meant to ward off predators while others draw in potential prey. Some behaviors are meant to impress mates while others help animals survive harsh conditions.

Internal mechanisms trigger some animal behaviors, such as body temperature causing an animal to move in or out of sunlight, while external factors trigger others, such as seasonal changes instigating bird migration. Studying animal behavior helps researchers make informed decisions about how to conserve species in the wild and in managed care settings.

The labs in this unit will help you understand how animals react and make adjustments in order to maximize their chances for survival and reproduction. From experimenting with bird courtship displays to constructing a beaver dam, you'll have the chance to explore the many things that animals *do* in order to succeed.

HIBERNATION HEART RATE

Investigate how deep rest affects heart rate, body temperature, and breathing.

PROCEDURE

1. Before you go to bed, place your lab notebook, thermometer, and timer set for 60 seconds next to your bed. You'll need these items in the morning! Now get some good sleep.

2. When you first wake up, take your temperature with the digital thermometer, and then count how many breaths you take as your timer counts down 60 seconds. Lastly, calculate your heart rate by counting how many beats you feel with two fingers on your neck beside your windpipe or on the thumb side of your inner wrist over 60 seconds. Record all three numbers in your lab notebook **(Fig. 1 and 2)**.

3. Once you are up and out of bed, set your timer for three minutes and start jumping rope! Try not to take any breaks, and just keep on jumping **(Fig. 3)**.

4. When time is up, re-record your body temperature, number of breaths in 60 seconds, and your heart rate. What effect did exercise have on your body? How does your exercised body differ from your resting body? Consider arranging your results into a simple table or colorful graph to share.

MATERIALS

→ Timer
→ Digital oral thermometer
→ Jump rope

SAFETY TIPS & HELPFUL HINTS

→ Your body returns to resting rates fairly quickly after you stop exercising. To get accurate data, consider having another person take your heart rate while you record your breathing.

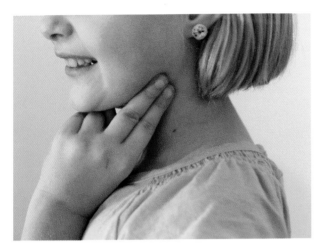

Fig. 1

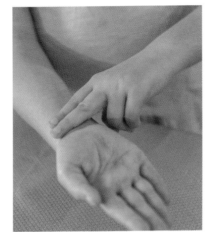

Fig. 2

Fig. 3

THE SCIENCE BEHIND THE FUN

→ Resting bodies use less energy than active ones, but some animals take energy conservation to a whole new level through hibernation, when they stay inactive for an extended period of time. Most animals that hibernate do so when food is scarce or temperatures drop (often these coincide, as in winter). Before hibernating, these animals gorge themselves on food to build up their fat reserves or store food to nibble on. Animals known to practice some form of hibernation include bears, rodents, bats, birds, amphibians, and some primates. In this lab, you'll investigate how rest and exercise affect body temperature, breathing, and heart rate.

CREATIVE ENRICHMENT

→ Consider investigating how deep rest effects other people's bodies, too. Have people you know try the experiment above so you can compare findings. You might also want to record their age, weight, and height to see if you can uncover any patterns in the data.

STANDING GUARD

Investigate whether sentry behavior helps protect the prize.

MATERIALS

→ Players (an even number, at least eight)

→ Large, safe outdoor area with varied terrain that has places to hide and run

→ Two delicious treasures, such as candy or cookies (in the traditional game, the prizes would be flags, but we're using sweet treats since potential predators are trying to capture and eat their prey) (Fig. 1)

Fig. 1

SAFETY TIPS & HELPFUL HINTS

→ Be sure to check for litter, sharp objects, and tripping hazards in the playing area before the rounds begin.

PROCEDURE

1. Split the players into two equal teams and head to your playing area.

2. Assign each team a territory. Make sure the two territories are separated by an open space.

3. Give each team a treasure to hide (don't let the other team see where!) and protect anywhere within their territory (Fig. 2).

4. Explain the game's three rounds.

Round One: *Seekers only.* All team members must leave their own territory and seek out the other team's treasure.

Round Two: *Seekers and Sentinels.* Half of each team seeks out the other team's treasure while the other half stays in their own territory to guard their own treasure.

Round Three: *Team Choice.* Each team decides amongst themselves how many Seekers and how many Sentinels to deploy.

5. During each round, Seekers tagged by the opposing team in that team's territory must return to their own territory. Sentinels must stay at least 10 feet (3 m) from their treasure unless an opposing Seeker is present. The round ends when Seekers successfully capture and return to their territory with the opposing team's treasure **(Fig. 3)**.

6. Record notes and thoughts about each round in your lab notebook. Did the first round end quickly? Did adding Sentinels make a round last longer? How did the teams decide to deploy Seekers and Sentinels in the third round? Did this affect the length of the game?

Fig. 2

Fig. 3

THE SCIENCE BEHIND THE FUN

→ Sentry behavior helps mammals and birds who live in groups survive. The sentry or sentinel stands guard, ready to warn the group of impending danger. It's easy to see this behavior in action when you observe meerkats or prairie dogs. One individual always stands upright, scanning the sky and landscape for would-be predators and other dangers. In this lab, you'll investigate the effectiveness of sentry behavior as you work alongside your team.

CREATIVE ENRICHMENT

→ Consider doing some research on other group-living mammal and bird species who employ sentinels, such as mongooses, babblers, scrub jays, and weavers. What kinds of special calls do they make? Is the same individual always the sentinel? Does more than one sentinel work at a time? There is a lot more to discover about this intriguing animal behavior.

MAKE AN IMPRESSION

Investigate the effects of elaborate courtship displays on those around you.

MATERIALS

→ Variety of dress-up supplies (colorful clothing, hats, wearable glitter, washable markers, non-toxic colored hairspray, etc.)
→ Variety of small gifts (tokens, candy, coins, shiny objects, paper statues, etc.)
→ Your own talents (singing and dancing)

SAFETY TIPS & HELPFUL HINTS

→ It's a good idea to plan what you'll sing and what dance you'll do *before* you're in front of anyone.

→ Make sure you only decorate yourself with non-toxic, washable stuff!

→ Be sure to ask for permission to use YouTube for the Creative Enrichment section of this lab. Also, be sure that Restricted Mode is enabled under Settings to help avoid inappropriate content.

PROCEDURE

1. Choose a morning when you know people who live with you will be around. The night before, gather and wrap several small gifts with bright colors and decorations **(Fig. 1)**.

2. When you get up the next morning, dress and behave as you normally would. Record how others react to your arrival.

3. Return to your room and change into the most colorful outfit you have. Decorate your skin, hair, and nails with washable markers, hair coloring, wearable glitter, and whatever else you have **(Fig. 2)**.

4. Leaving your room with your gifts in hand, sing loudly and dance flamboyantly for everyone you see. Hand them gifts. Again, record their reactions to your arrival.

5. If you're feeling super brave, wear the outfit and display the song and dance behaviors for your friends to see their reactions.

Fig. 1

Fig. 2

→ Individual animals go to great lengths to find and attract a mate or mates. Some of the most interesting strategies involve elaborate presentations of singing, dancing, decoration, and gift-giving, also called courtship displays. Impressive animal courtship displays are all around us, from spiders to frogs to whales, but are, by far, most striking in birds. For example, the males of many species of bowerbirds and birds of paradise construct elegant stages on which to perform intricate dances that seem to mesmerize females with color, sound, movement, and treats. But these incredible displays, with their bright coloration and loud vocalizations, don't just cost a lot of time and energy to produce. They put the performer at an increased risk of predation, which further demonstrates the great pressure animals are under to reproduce. In this lab, you'll document the effects of colorful "courtship" displays on your family and friends.

CREATIVE ENRICHMENT

→ Nature documentarians have captured elegant and enchanting animal courtship displays that are simply amazing to watch! With permission, search YouTube for courtship displays of bowerbirds, peacock spiders, birds of paradise, manakins, alligators, and peacocks. Record your thoughts about each video segment. Were the performers more often male or were they female? What colors were most commonly displayed?

SILK HUNTERS

Discover the incredible engineering skills of spiders around your home.

SAFETY TIPS & HELPFUL HINTS

→ Although spiders can be venomous, most species are harmless. They are also vital to controlling pest insect populations (flies, mosquitoes), so observe them as part of this lab, but do not harm them.

→ It's often easier to see spider webs in the early morning when dew is still on the threads.

PROCEDURE

1. Find as many spiderwebs as you can inside and outside. Whenever you find one, carefully make a sketch of it in your lab notebook and then photograph the web.

2. Record thoughts and notes about each web. Ask yourself what type of web it is (sheet, orb, tangle/cob, funnel, tubular, etc.). What is the web attached to? Could you find the spider? Is prey caught in the web? Is the web dirty or clean? What type of web is most common in this area?

CREATIVE ENRICHMENT

→ Consider creating a mini museum exhibit about spiders and their use of silk. Do some online research or visit your school library to discover additional uses of spider silk, beyond web-building. Display the drawings from your lab notebook and the digital photos that you captured in order to share what you have learned about different web types and the body parts involved in web construction. Spiders are often underappreciated, so do your best to turn your friends and family into spider champions!

MATERIALS

→ Handheld magnifying glass
→ Digital camera or handheld device with camera (smartphone, tablet, etc.)

THE
SCIENCE
BEHIND THE FUN

→ Webs are one of the most defining features of spiders. Scientists believe that spiders first produced silk from their spinnerets to protect their eggs and bodies, later evolving to create webs for hunting. All spiders can make silk, but not all build webs. Those that do usually use one of a few common types:

Sheet webs are flat, horizontal, and usually found on the tops of hedges **(Fig. A)**.

Orb webs are circular, vertical, and usually hang high above the ground **(Fig. B)**.

Tangle webs (also called cobwebs) look like a messy jumble of threads **(Fig. C)**.

Tubular webs are usually found on the ground **(Fig. D)**.

Funnel webs allow the spider to hide at one end and run out when they detect prey **(Fig. E)**.

Some spiders use their silk for hunting, but, rather than building a web, they throw it at their prey, either in a silk net or a long line with a sticky end. Still others make trap doors out of silk and small particles in their environment. These spiders hide in a tunnel they've dug below the door, which has silk hinges, and jump out to grab their prey. In this lab, you'll document the web types found around your home.

Fig. A

Fig. B

Fig. C

Fig. D

Fig. E

DESERT SHADE LOVERS

Discover how desert animals use shade and burrows to survive the heat.

MATERIALS

→ Digital outdoor thermometer
→ Timer
→ Hand shovel/trowel

SAFETY TIPS & HELPFUL HINTS

→ You'll need to conduct this lab on a sunny day.

→ It can be dangerous to let yourself get too hot, so use an outdoor digital thermometer for this lab, instead of your own body.

→ Get permission to dig before you start!

→ Tying string to the thermometer will make it easier to retrieve when you measure the temperature in the burrow.

PROCEDURE

1. Draw a table with three columns and twenty rows in your lab notebook. Label the columns sun, shade, and burrow.

2. Dig an artificial burrow outside in a sunny spot. Keep in mind that burrows usually run horizontally under the ground. You can start the process with a hole, but you'll need to then dig horizontally so sunlight does not shine into the end of the burrow (this is the protective area for a hiding desert animal) **(Fig. 1)**.

3. Set the timer for 30 seconds, then record the outside air temperature in your lab notebook. Place the thermometer in direct sunlight (and you in nearby shade) and record the temperature every 30 seconds for 10 minutes in the sun column of your lab notebook **(Fig. 2)**.

4. Once you record 10 minutes of readings, cool the thermometer back to your starting temperature.

5. Repeat the experiment, this time with the thermometer in the shade. Record the temperature every 30 seconds for 10 minutes in the shade column of your lab notebook, then cool the thermometer back to your starting temperature again.

6. For the third and final trial, place the digital thermometer as far into your artificial burrow as possible. Record the temperature every 30 seconds for 10 minutes in the burrow column of your lab notebook **(Fig. 3)**.

7. Compare your temperature readings. Was the shade cooler or warmer than the burrow? How much hotter was it in the sun?

Fig. 1

Fig. 2

Fig. 3

THE SCIENCE BEHIND THE FUN

→ In the last unit, you saw how lighter colors of fur, feathers, scales, or skin can help some animals survive in desert heat. Behaviors can also increase survival there. Many desert animals (coyotes, rabbits, and songbirds) are crepuscular, meaning they are only active at dusk and dawn. Nocturnal animals (foxes, owls, and many snakes and rodents) are most active at night, when temperatures are cooler. Some desert animals (tortoises, toads, and some squirrels) even enter a state of dormancy similar to hibernation during the hottest parts of the year. In this lab, you'll document the heat intensity of life in direct sunlight, in the shade, and in an underground burrow.

CREATIVE ENRICHMENT

→ Investigate the effect of burrow design and dynamics for limiting heat exposure. Sketch three different burrow designs in your lab notebook, then start digging. Record the temperatures of each using the above lab procedure to determine the most effective design.

BEAVER BUILDERS

Build an effective dam like a dedicated beaver parent.

→ Beavers don't wait until the riverbed runs dry to construct their dams, so neither should you (that's cheating!). Be sure the water is running while you try out your various designs and materials.

PROCEDURE

1. Dig a small river system with a slight incline (so the water won't pool on its own) **(Fig. 2)**.

2. With the hose at the top of the river, turn on a slight flow of water (you don't want roaring rapids).

3. Use your gathered natural materials to build a dam in the middle of the river system. The point is to get water to pool behind the dam, thereby creating the pond habitat that a beaver would use to build its lodge. Try different configurations, different building materials, dry dirt versus wet mud, etc. **(Fig. 3)**.

4. Draw the various designs in your lab notebook and label the materials used in each. Record your thoughts on which designs and materials worked best—and which didn't work at all!

MATERIALS

→ Outdoor space that can get dirty
→ Small shovel
→ Hose hooked up to a water source
→ As much natural building material (sticks, twigs, dirt, rocks, sand, etc.) as you can find

SAFETY TIPS & HELPFUL HINT

→ If you don't have a nearby space to dig in the dirt, then you can construct an artificial riverbed out of aluminum foil **(Fig. 1)**.

Fig. 1

Fig. 2

Fig. 3

THE
SCIENCE
BEHIND THE FUN

→ Whether they build an underground burrow, a nest along the beach, or a hive in a tree, many animals are experts at finding and using materials to create a safe space to live and raise young. The beaver is one of the most skilled animal architects, constructing both a dam and a lodge as it starts its journey into parenthood. Beavers use their sharp teeth and powerful jaws to gnaw through tree trunks before dragging them into the water and cementing them together with mud. Once the dam is strong enough to hold back water, the beaver uses the newly created pond for its lodge-building site. The lodge is hollow inside, with a floor of softly shredded wood. Its underwater entrances protect baby and adult beavers alike from would-be predators. In this lab, you'll try your hand at dam construction to see which materials and designs work best.

CREATIVE ENRICHMENT

→ Constructing an effective dam is only half the job for a dedicated beaver parent. Building a dry, warm den inside of a lodge is even more important, since this is the nursery for raising beaver kits. Once you build the perfect dam, consider building a lodge structure in the pond that is created behind it. The goal is to construct a hollow, dry space that is protected from the outside world. Use materials of your choice. Once completed, try placing a dry cotton ball inside the den to see if your beaver kits would stay safe and warm.

PLAYING TRICKS

Explore how clever tricks help animals survive.

MATERIALS

→ Group of friends who like playing games
→ Squirt bottle filled with water

SAFETY TIPS & HELPFUL HINTS

→ Before you play any tricks, remember the pledge that you took at the beginning of this book to **DO NO HARM.** While the tricks in this lab are meant to be fun and informative, be absolutely sure not to put anyone in danger.

→ In the third game scenario, use only clean water with nothing else added. A horned lizard uses acid-laden blood to ward off predators, but you should only use fresh drinking water!

→ Ask for permission to use YouTube for the Creative Enrichment section of this lab. Also, be sure to enable Restricted Mode under Settings to avoid any inappropriate content.

PROCEDURE

1. Scenario one: You are a killdeer trying to protect your chicks from a predator. Invite your friends to play tag, but don't volunteer to be it. When the game starts, pretend to hurt your foot and limp around so the tagger will run after you instead of your brood of chicks (the other friends playing). Just before the tagger reaches you, drop the act and run away as fast as you can. If you employ this trick effectively, your brood of chicks (your other friends) will escape while the tagger chases you, just like a mama killdeer pretends to be injured to lure away a predator **(Fig. 1).**

2. Scenario two: You are a stealthy ambush hunter trying to attract your prey by mimicking the sounds of a baby animal. Invite your friends to play hide and seek, but don't volunteer to be it. Once you're in your hiding place, make soft baby animal noises, such as "peep" or "eeh," every few seconds to attract the seeker. Once they are close, jump out and grab them! If you employ this trick effectively, you will get a tasty meal, just like a margay attracts neighboring primates by mimicking the sounds of a baby tamarin **(Fig. 2).**

3. Scenario three: You are a horned lizard trying to ward off a chasing predator. Invite your friends to play tag, but don't volunteer to be it. Hide a squirt bottle of water in your pocket or shirt. When the game starts and the tagger gets close to you, spray them in the face with water and run as fast as you can to get away. If you employ this trick effectively, you'll be able to evade the tagger, just like a horned lizard squirts blood from its eyes to escape a predator **(Fig. 3).**

Fig. 1

Fig. 2

4. Record your friends' reactions to each trick you tried. Do you think animals who use these tricks get the same reaction? How might it be different?

Fig. 3

→ Many creatures in the animal kingdom use clever tricks to increase their chances of survival. Some animals pretend to be injured or dead in order to confuse would-be predators while others release parts of their bodies as a diversion— some lizards, for instance, can do this with their tails (which they can then regrow!). Some animals shoot out body fluids, such as ink or blood, to make a quick escape. Others mimic the sounds of different animals in order to bring prey closer, ward off predators, attract mates, or steal food items by scaring others away. In this lab, you'll try out a few of these clever tricks to see what effect they have on those around you.

CREATIVE ENRICHMENT
→ Nature documentarians have recorded many of the clever tricks that animals play, and these can be amazing to watch! With permission, search YouTube for superb lyrebird, horned lizard blood squirt, killdeer broken wing act, cuttlefish ink discharge, and burrowing owl rattlesnake call. Record what you learn from each video and think up other fun games to play based on these unique animal behaviors.

FASCINATING ANIMAL
SENSES

We can celebrate that humans as a species can see, hear, smell, taste, and touch, but frankly, our senses pale in comparison to those of other animals. Night vision, ultraviolet detection, infrared sensing, echolocation, electricity sensing, magnetic orientation, infrasound communication, long-distance odor detection—all of these extraordinary sensory abilities (and more!) allow species to thrive.

The labs in this unit will help you better understand how animals use their senses and associated organs to perceive the world around them. From trying your hand at echolocation to documenting the perception skills of your own fingers, this unit gives you the chance to detect and respond to stimuli the way other animals do.

SEEING WITH SOUND

Explore the role of feedback in bat echolocation.

Fig. 1

PROCEDURE

1. Use your measuring tape and sidewalk chalk to mark off 1-foot (0.3 m) intervals on the ground from the wall outward up to 10 feet (3 m) (a total of 10 marks) **(Fig. 1)**.

2. Starting 10 feet (3 m) away from the wall, throw your ball at the wall and have someone record how many seconds it takes for the ball to leave your hand and return back to your hand. Using the same force of throw (as best you can), repeat this procedure for each distance interval leading up to the wall **(Fig. 2)**.

3. Graph your results with seconds on one axis and distance on the other. If the ball is you (a bat) emitting a click, what do these results tell you about how the feedback from an object (the wall) relates to its distance from the sound source? Some bats can emit up to 200 clicks *per second*—just think about that for a moment! These little mammals are absolutely incredible.

MATERIAL

→ Tennis ball or small rubber bouncy ball
→ Tape measure or meter tape (least 10 feet [3 m] long)
→ Sidewalk chalk
→ Durable outdoor wall with no protrusions or windows
→ Timer
→ Someone to help you keep time

SAFETY TIPS & HELPFUL HINTS

→ Do your best to throw the ball with the same strength and at the same speed for each distance interval.

Fig. 2

THE
SCIENCE
BEHIND THE FUN

→ Interestingly, not all animals use eyes to see the world around them. Some species use biological sonar, or echolocation, to sense the presence and position of objects in their environment. They do this by emitting sounds and receiving feedback in the form of echoes that bounce off nearby objects. Varying forms of echolocation play a critical role in navigation and hunting for a variety of species, including bats, some whales and dolphins, shrew species, and even some birds. Echolocation allows bats to make use of the nighttime hours in ways that other animals cannot. Because many insects only come out at night and many insectivores (insect eaters) do not, hunting at night gives bats an abundant food source with less competition. It also reduces the risk of bats themselves being eaten! Bats emit a series of rapid clicks to avoid obstacles and zero in on their prey. In this lab, you'll investigate the role of feedback in bat echolocation through a modified game of Wall Ball.

CREATIVE
ENRICHMENT

→ This activity helped you explore echolocation by having you send out and receive signals from different distances. But bats and other echolocating animals don't just receive echoes from stationary objects that are directly in front of them. Their two ears, situated apart from each other, receive feedback from all around. Get some friends together outside in a safe open space for a game of Marco Polo. Put on a blindfold and try to tag your friends as you call "Marco" and they immediately respond "Polo." Not an easy task, but certainly made simpler with two functioning ears.

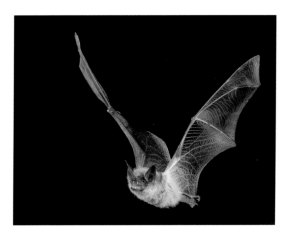

LAB
22

ON THE SCENT TRAIL

Investigate how different scents affect ant behavior and signaling.

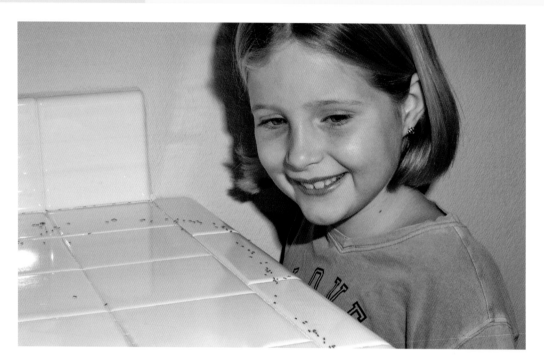

MATERIALS

→ Fresh cotton balls
→ Variety of sweet/pleasant scent samples (sugar water, non-citrus fruit juice, milk, soda, etc.) (Fig. 1)
→ Variety of bitter/spicy scent samples (hot sauce, vinegar, soapy water, citrus juice, coffee, rubbing alcohol, etc.) (Fig. 2)
→ Timer

SAFETY TIPS & HELPFUL HINTS

→ Be sure *not* to hurt any ants during the course of this lab. Work around them carefully and do not submerge them in the scent samples.

→ Scout out some good ant habitat before starting the lab so you don't spend all your time searching for ants. They are likely to be most active in the morning or late afternoon hours, not during the hottest parts of the day.

Fig. 1 (clockwise from top left: orange sports drink, sugar water, apple juice, milk)

Fig. 2 (clockwise from top left: coffee, lemon juice, vinegar, orange juice)

PROCEDURE

1. Head outside and find an active ant trail (or you can also work with ants inside your home if they happen to be present).

2. Saturate a cotton ball with your first testable scent, place it about 3 inches (7.6 cm) from the trail, and set your timer for 3 minutes. Observe the reaction to each scent for 3 minutes, then carefully remove the cotton ball and gently brush the ants onto the ground. If the trail is long enough, consider testing each scent on a different section of the trail, so you are engaging different ants with each trial **(Fig. 3)**.

3. Record the ants' reactions to each cotton ball in your lab notebook. Do the ants take time to investigate every cotton ball or do they ignore some altogether? Do they respond differently to the positive (sweet/

Fig. 3

pleasant) and negative (bitter/spicy) scents? After they initially investigate the negative scents, do they leave them alone and signal others to stay away? Which scents keep their attention the longest? How many individual ants swarmed onto each cotton ball?

4. Consider turning these interesting data into colorful graphs or charts to share.

→ Most people know that ants have an amazing ability to carry up to 50 times their own body weight, but they also have another impressive power: their extraordinary sense of smell! They process chemical signals called pheromones using their antennae, allowing them to distinguish colony members from foes, send alarm signals to their neighbors, and alert the colony to food resources. While scientists have known for many years about ants' use of positive pheromones to lead each other to food, they've only recently documented the use of negative pheromones to tell fellow colony members where *not* to go. In this lab, you'll examine the effect of different scents on the signaling behavior of ants.

CREATIVE ENRICHMENT

→ It was interesting to observe the different reactions to positive and negative scents, but it might also be interesting to observe with different concentrations of those scents. Dilute each scent sample with water and try the experiment again. Were ants' reactions similar to when the scents were more concentrated? Were they more tolerant of the negative scents when they were diluted? Record your findings for further comparison.

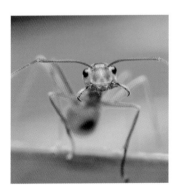

LAB 23

GOOD VIBRATIONS

Experiment with sending vibrations through different surfaces.

Fig. 1

MATERIALS

→ Rubber mallet (Fig. 1)
→ Tape measure
→ Sidewalk chalk (for outdoor surfaces) or painter's tape (for indoor surfaces)
→ Ear plugs or headphones
→ Several different floor/ground surfaces
→ Partner

SAFETY TIPS & HELPFUL HINTS

→ Do your best to smack the rubber mallet at each distance interval with the same force.

→ It should go without saying, but be careful when smacking surfaces with the rubber mallet. Do not damage any property or get yourself in trouble.

PROCEDURE

1. Identify several floor and ground surfaces inside and outside of your house that are made of various materials—wood, tile, dirt, stone, carpet, vinyl, etc.

2. Draw a table with five columns in your lab notebook. Label them 5, 4, 3, 2, 1. Label a row for each surface you'll use.

3. On each surface (do one at a time), measure out 5 feet (1.5 m) and place a mark at each foot in a straight line using chalk or painter's tape (five marks total) **(Fig. 2)**.

4. With the other person sitting or lying at one end of the 5-foot (1.5 m) line, smack the rubber mallet onto the surface at the opposite end. (NOTE: Encourage your partner to close their

Fig. 2

eyes and wear earplugs or headphones so they can concentrate on the vibration rather than sight and sound). Record a 1 if they did not feel it, 2 if they felt it a little bit, or a 3 if they felt it a lot. Repeat the process at each marked interval until you are 1 foot (0.3 m) away, recording each reaction in the appropriate column **(Fig. 3)**.

5. Repeat this procedure for at least two more surface types, then graph each one with distance on one axis and reaction on the other. Which surfaces carried vibrations most effectively? Did any surfaces transmit no vibrations at all?

Fig. 3

→ Many animals hear much better than humans, detecting sounds far outside the range of human hearing. But not all perceive sound the same *way* we do. Several species hear and communicate through vibrations in surfaces where they live. For example, elephants can send and receive messages over great distances through ground vibrations. Snakes have special bones in their jaws that allow them to detect and interpret vibrations from nearby prey items. Some spiders sneakily send vibrations through the webs of other spiders to get them to come closer while others drum their appendages to get the attention of potential mates. But transmission surface makes a big difference in how sound vibrations travel and how they are detected by their intended recipients. Think about a treehopper signaling with vibrations on a flower stem, an alligator attracting mates with vibrations through the water, and a kangaroo rat drumming in the sand underground. In this lab, you'll investigate how vibrations travel differently through a variety of surfaces.

CREATIVE ENRICHMENT

→ Consider testing how well vibrations travel through other surfaces as well, including couches, pillows, tables, doors, glass, metal, etc. If the surfaces are fragile, use your finger rather than the rubber mallet!

FEELING FOR FOOD

Examine how reliable the human sense of touch is for identifying food items.

MATERIALS

→ Blindfold
→ Variety of small food items (apple, gummy bear, pretzel, assorted chips, assorted crackers, cheese chunks, peas, blueberries, etc.)
→ Variety of small, non-toxic non-food items (playdough pieces, assorted toys, erasers, toy food from play kitchen set, bar of soap, etc.)
→ Participants

SAFETY TIPS & HELPFUL HINTS

→ **Don't use any toxic or dangerous items for this lab!** Make sure you ask about allergies, too.

→ Be sure participants don't cheat by using their senses of sight or smell to test the potential food items. They must rely *only* on their sense of touch to identify each item as food or not. The whole idea is to see if you can trick them by choosing items that are hard to discern as edible.

→ Test one person at a time, and be sure no one else is watching before their turn.

PROCEDURE

1. Blindfold your first subject and have them agree to use only their sense of touch to decide if each item is food or not (they will likely need to hold the items far away from their face to avoid smelling) **(Fig. 1)**.

2. Place the first object in your subject's hands. Let them feel it and move it around in their hands until they decide whether or not it is edible. If they decide that the item is edible, ask them to identify the kind of food item it is and to put it in their mouth for a snack.

3. Record data for each subject in your lab notebook. Were some people better than others at deciphering food from non-food? Did anybody cheat and use their sense of smell? What was toughest to discern? Which items were easiest?

4. Consider turning these interesting data into colorful graphs or charts to share.

Fig. 1

→ While most animals rely on sight, sound, smell, and even taste to find and capture their food, a few animals rely instead on their sense of touch. Animals that live in dark environments, such as murky water or underground, have honed their sense of touch to find the food they need. These animals often have adapted appendages that help them forage successfully. The feelers on the star-nosed mole, for example, are far more sensitive to touch than the human hand, allowing this little creature to locate, identify, and capture tiny invertebrates with incredible speed. In this lab, you'll document the human hand's effectiveness in identifying food.

CREATIVE ENRICHMENT

→ Lab 22 had you explore sense of smell in ants, but here you'll explore sense of smell in humans. Once you record your data on human sense of touch, consider testing the effectiveness of smell among your participants for identifying food objects. You will probably want to use a brand-new assortment of food items for this activity, so your test subjects won't already know which items might be in play. Be sure to record data for each person you test and look for patterns in your results.

TASTE TESTERS

Explore how your sense of taste stacks up against that of others.

MATERIALS

→ Sugar water (sweetness/sugars)

→ Saltwater (saltiness/salts)

→ Tonic water (bitterness/ bittering agents)

→ Lemon juice (sourness/acids)

→ Chicken or beef broth (savory/meat or cheese flavor)

→ Five clean eye droppers

→ Taste-testers

SAFETY TIPS & HELPFUL HINTS

→ Be prepared for some testers not to like what they are tasting. Be sure that none of them have allergies to any of the taste solutions before you begin. Thank all testers for their willingness to help you collect this data.

→ Record your recipes (1 tablespoon [13 g] sugar with 1 cup [235 ml] water, lemon juice straight from the lemon, etc.) in your lab notebook so you know the concentration.

PROCEDURE

1. Put each of your five taste characteristic solutions in separate cups and label them carefully so you don't mix them up **(Fig. 1)**.

2. Have your testers try each solution and record their reactions in your lab notebook. Did they like the taste? Which of the five taste characteristics do they think the solution represents? What other foods or drinks do they think might have this same taste characteristic? Be sure to test and record results for yourself too **(Fig. 2)**!

3. Consider turning your data into colorful graphs or charts to share.

CREATIVE ENRICHMENT

→ Based on your data, it should be evident that human taste receptors do not produce the same results across different people. In addition to testing each person's reaction to the five taste characteristics, it is also interesting to examine the sensitivity of each person's ability to sense the characteristics. Consider making three different dilutions of each taste solution and run your taste trials again (keep one cup of each solution the same as the initial trial, but then add a more concentrated version and a more dilute version of each solution). Are the younger people in your taster group more sensitive to less concentrated solutions? What other patterns did you notice when testing the different dilutions?

lemon juice   tonic water   sugar water

salt water   chicken broth

Fig. 1

Fig. 2

THE
SCIENCE
BEHIND THE FUN

→ Creatures across the animal kingdom rely on taste receptors, often in combination with scent receptors and visual cues, to identify safe and nutritious foods. While human taste buds are located primarily on the tongue, many other animals have their taste receptors located on vastly different areas of their bodies. The taste receptors of most vertebrate animals are bundled in the mouth (though fish also have them on the outside of their bodies). The location of taste receptors in invertebrate animals varies widely: octopuses have them on their suckers, and insects may have them on their legs, antennae, feet, and/or jaws!

Animals also differ in the number of taste receptors they possess, as well as the taste characteristics they can sense. Most humans can sense five different taste characteristics: sweet, sour, bitter, salty, and savory (also called umami). And most humans have approximately 10,000 taste receptors, but they don't work the same way in every person. This is what accounts for taste preferences. In fascinating contrast, catfish have more than 100,000 taste receptors located all over the *outside* of their bodies but primarily bundled on their super-sensitive whiskers, also called barbels. In this lab, you'll see how your own sense of taste differs from that of your test group.

TERRIFIC TONGUES

Take a closer look at how important a cat's tongue is to its daily routine.

PROCEDURE

1. Put a drop of tuna water on your index finger and have a friendly cat lick it off. Record your thoughts about how that felt. You can also wait until the cat is grooming and simply insert your fingers into the grooming space so the cat licks you instead of its fur. After washing your hands, lick your own finger to experience the difference **(Fig. 1)**.

2. It would be awfully gross to lick your arms and legs to groom like a cat, so instead, try drinking like a cat. Pour a ½ cup (120 ml) of tuna water into a bowl on the kitchen floor and use your timer to record how long it takes the cat to drink all the liquid. Consider using your digital camera or other handheld device to record the cat drinking, so you can review it in slow motion **(Fig. 2)**.

3. Now pour the same amount of your favorite drink into a clean bowl and try drinking it like a cat (slap your tongue onto the surface, then pull straight up, closing your mouth around the column that is created) **(Fig. 3)**.

MATERIALS

→ Friendly cat
→ Two wide-mouthed bowls
→ Water drained from can of tuna fish
→ Your favorite drink
→ Timer
→ Digital camera or handheld device with camera (smartphone, tablet, etc.)

SAFETY TIPS & HELPFUL HINTS

→ If you don't have ready access to a friendly cat, consider visiting someone who does.

→ If you are allergic to cats, consider acting as an observer while a friend or family member completes this activity.

Fig. 1

Fig. 2

Fig. 3

THE SCIENCE
BEHIND THE FUN

→ It turns out that, in the animal world, tongues have far more functions than just identifying safe and nutritious food items. Animals use their tongues for all sorts of things, and the actual shape, size, and appearance of tongues varies widely across species. Some animals use their tongues to generate words (humans), others to attract prey (alligator, snapping turtle), some use them to capture prey (chameleon), some use them to cool their bodies (dog), and some animals even use their tongues to smell the air (snake). Cats have particularly interesting tongues that are covered with tiny spines, called papillae, which they use to clean their bodies as well as eat and drink—actually, the papillae of some big cat species can even scrape meat off the bones of their prey! Researchers have taken a closer look at how cats use their tongues for drinking and have discovered remarkable things. Unlike the lapping up motion of dogs, cats gently slap the surface of the liquid, which creates a column of liquid that they close their mouths around. Amazingly, some cats can do this slapping motion up to four times a second! In this lab, you'll see how your tongue stacks up to that of a cat.

CREATIVE ENRICHMENT

→ Consider putting yourself in the role of other interesting species to see how they use their tongues. Stick the tip of your tongue to small food items like a chameleon does, pant after exercising to cool off like a dog, scare your friends away like a blue-tongued skink by temporarily staining your tongue with a popsicle, and stick your tongue out at dinnertime to taste the air around you like a snake. Or maybe imagine what life would be like if you couldn't move your tongue. The tongues of crocodiles are firmly attached to the bottoms of their mouths, for example. Try singing *Twinkle, Twinkle, Little Star* with your tongue planted firmly against the bottom of your mouth. And try communicating with your friends and family without the use of your tongue. This is one important sense organ!

LAB 27

EXTRAORDINARY SENSES

Experiment with simple devices that enhance your vision.

PROCEDURE

1. Hang an open book on a wall of a large indoor room or outside, where you can stand several feet away (if there's nothing on which to hang the book, have someone hold it for you). Measure out 10 feet (3 m) from the book and try to read the words on the open pages out loud. Record your thoughts on the experience in your lab notebook (**Fig. 1**).

2. Now, stand the same distance away and use the binoculars to try reading the open pages again. Record your thoughts on the experience in your lab notebook (**Fig. 2**).

3. Observing each household object from about 6 inches (15 cm) away, draw in your lab notebook what you see. Label any interesting parts and take notes about any noticeable features (**Fig. 3**).

4. Now, use the handheld magnifying glass to re-examine each of the objects and draw new pictures from your enhanced view. Remember to label any interesting parts and record your thoughts on noticeable features (**Fig. 4**).

MATERIALS

→ Binoculars
→ Handheld magnifying glass
→ Paperback books with words
→ Small household objects (salt granules, pepper granules, ice cube, strands of hair, short piece of yarn, feather, etc.)
→ A tape measure or meter tape (at least 10 feet [3 m] long)

SAFETY TIPS & HELPFUL HINTS

→ Online retailers offer many affordable sense-enhancing devices (night vision goggles, UV lamp/ black light, bionic listening device, telescope, microscope, etc.) that can extend this investigation.

Fig. 1

Fig. 2

Fig. 3

Fig. 4

THE SCIENCE BEHIND THE FUN

→ Scientists and engineers have developed an array of incredible inventions to enhance human senses so they can discover and better understand the environment around them. Many of these inventions are based on lessons from other species. Think about binoculars, telescopes, microscopes, bionic listening devices, night vision goggles, UV lamps, infrared detection systems, x-ray machines, carbon monoxide detectors, and radar/sonar. In this lab, you'll experiment with two common devices designed to enhance your sense of vision (one for faraway objects and one for close-up objects).

CREATIVE ENRICHMENT

→ While these two simple devices allow you to experiment with vision enhancement at home, many other complex sense-enhancing devices are used every day throughout the world. Consider borrowing a pair of night vision goggles, a UV lamp, a bionic listening device, a telescope, or a microscope (or save up to purchase your own). Experiment with these devices to see how they enable you to see and hear beyond the normal human range. Consider doing some research at your school library or online to discover other devices and techniques that scientists and engineers have developed to learn more about the world, such as x-ray technology, thermal imaging, near infrared spectroscopy, virtual reality gloves, wristband heart monitors, and augmented reality devices.

AMAZING ANIMAL MOVEMENT

It is amazing to discover the incredible diversity of habitats around which animals build their lives, taking advantage of some of the harshest and most remote environments on the planet. But in order to be successful, their bodies must be well-adapted for movement in those spaces. Several arboreal species swing expertly through the trees, aerial animals may fly effortlessly through the air, many aquatic and semi-aquatic animals swim smoothly through the water, fossorial animals dig easily underground, and many terrestrial species run agilely on land.

From fins to claws to wings, animals use their bones, musculature, and other bodily structures to catch prey, outrun predators, and gain access to unique resources in the niches where they live.

The labs in this unit will help you discover how animals use specialized body features to move efficiently through their environments. From jumping like a frog to swimming like a sea lion, this unit gives you the chance to compare your sports skills to some of the animal kingdom's most notable athletes.

WALKING THE WALK

See how well your body is adapted for walking like a human.

MATERIALS

→ Timer

SAFETY TIPS & HELPFUL HINTS

→ Give yourself a short rest in between each movement style so your body and muscles can reset. This is also a good time to record your thoughts about each experience.

→ You may want to watch some short video clips before imitating each movement style so you can get a better feel for how each animal moves in its environment.

PROCEDURE

1. Draw a data table in your lab notebook with two columns and five rows. In the first column, write the names of the five animals you'll be imitating: crab, gorilla, snake, frog, and human. In the second column, you'll be recording your thoughts on how each movement style felt on your body and your ability to move over short distances **(Fig. 1)**.

2. Set your timer for one minute, and start the countdown as soon as you begin to walk like each animal.

Crab: Squat down low, hold your arms out in front of your body, and walk sideways quickly, first to the left, then to the right. Continue this movement style for one minute **(Fig. 2)**.

Gorilla: Close your hands into tight fists and put them down on the floor. Walk around on your fists and feet for one minute **(Fig. 3)**.

Snake: Lie down on your belly with your arms by your sides and your legs close together. Slither around for one minute **(Fig. 4)**.

Frog: Get down on your hands and feet. Hop around for one minute **(Fig. 5)**.

Human: Walk around normally for one minute.

3. Which parts of your body were uncomfortable after using each movement style for one minute? How long do you think you could have lasted using each style? What aspects of your body make it well-adapted for walking in an upright position?

Fig. 1

Fig. 2

Fig. 3

Fig. 4

Fig. 5

THE SCIENCE BEHIND THE FUN

→ Animal life began in the oceans millions of years ago, but many modern-day species build their lives in terrestrial habitats. Terrestrial species exhibit a whole host of locomotion styles, many requiring specialized body shapes and appendages. Humans, for instance, walk in a bipedal position; fully upright movement using primarily the legs and feet. In this lab, you'll investigate the movement styles of some common terrestrial animals and discover which best suits your body.

CREATIVE ENRICHMENT

→ Have you ever considered what is actually happening with your body when you walk? Ask someone to walk slowly back and forth or record a short video of yourself walking and then watch it. What parts of the body are moving and when? What does each foot do? What joints are involved? Do the arms move also? Are both feet ever touching the ground during the same time? Do the feet roll or stomp on the ground? What muscles are important for moving the legs forward? Bipedal walking is actually quite complicated and rare in the animal world!

LET'S JUMP!

Find out how your leaping skills stack up against the animal kingdom's best jumpers.

SAFETY TIPS & HELPFUL HINTS

→ The best way to compare your jumping skills to that of other animals is to consider jumping distance (horizontal) and height (vertical) in relation to overall body size. To do this, measure your own height (bottom of the feet to the top of the head—stand up straight!) and width (back to front from a side view) and record the distance and height of your best jump attempts **(Fig. 1 and 2)**.

→ Be sure to run your trials on a safe jumping surface, ideally soft grass or sand for jumping distance and a soft mat for jumping height.

→ It's best to have someone help you measure the distance and height of your jump attempts so you can fully focus on going as far and as high as possible.

continued on page 80

MATERIALS

→ Tape measure or meter tape (at least 10 feet [3 m] long)
→ Short piece of rope
→ Yardstick
→ Wooden stick
→ Assistant

Fig. 1

Fig. 2

THE SCIENCE BEHIND THE FUN

→ Many species in the animal kingdom have incredible jumping skills, in terms of how high and how far they can jump. Some are good at one or the other, but some are good at both. Unique body shapes and musculature, often found in prey species who need to escape predators, make this outstanding jumping possible. Some of the world's best jumpers include various kangaroo species (nearly 7 times their own length), klipspringers (more than 10 times their own height), rabbits (more than 15 times their own length), kangaroo rats (nearly 30 times their own length), grasshoppers (up to 30 times their own length), frogs (more than 40 times their own height), jumping spiders (up to 50 times their own length), fleas (nearly 50 times their own length), and froghoppers (more than 70 times their own height). In this lab, you'll calculate your jumping distance and height to see how you compare to other species.

CREATIVE ENRICHMENT

→ Now that you have a better sense of your own jumping skills, use this lab to calculate jumping distance and height for your friends and family. What patterns can you uncover in jumper body shape, size, and age? Another important factor is weight, so consider recording this important data point for your jumpers and see if you can uncover additional patterns.

Fig. 3

Fig. 4

PROCEDURE

1. Stretch out 10 feet (3 m) of your measuring tape on the jumping surface. Lay the rope perpendicular to the beginning of your tape to act as a starting line. Have your assistant stand alongside the tape measure so they can clearly see where you land **(Fig. 3)**.

2. Standing long jump. Stand with your toes up against the start line and jump as far forward as you can. Repeat the jump three times and record the longest distance in your lab notebook **(Fig. 4)**.

3. Running long jump. Start running as fast as you can from approximately 10 feet (3 m) back from the start line. When you reach it—don't step past it!—jump as far forward as you can. Repeat the jump three times and record the longest distance in your lab notebook **(Fig. 5)**.

4. Have your assistant hold the yardstick vertically and hold the wooden stick horizontally fairly low beside it.

5. Standing high jump. Stand in place beside your assistant and jump over the stick. Have your assistant raise it higher, and then jump over the stick again. Keep raising the stick until you can no longer jump over it. Record your highest jump in your lab notebook **(Fig. 6)**.

6. Running high jump. Start running as fast as you can from approximately 10 feet (3 m) back from your assistant, and jump over the stick when you reach it. Have your assistant raise the stick higher, and then repeat the jump. Keep raising the stick until you can no longer jump over it. Record your highest jump in your lab notebook.

7. Time to analyze your results! To calculate how far you jumped in body lengths, divide the distance that you jumped by the width of your body (make sure you use the same unit of measure—inches, centimeters, etc.—for both!). Similarly, to calculate how high you jumped in body heights, simply divide the height that you jumped by your own height (again, use the same units of measure). So, how do you compare to other jumpers in the animal kingdom!?

Fig. 5

Fig. 6

CHEETAH CHALLENGER

Discover how your sprinting abilities compare to those of a cheetah.

MATERIALS

→ Timer
→ 100-meter (110-ft) measuring tape
→ Obstacles (traffic cones, boxes, stuffed animals, pillows, etc.)
→ Friend(s)

SAFETY TIPS & HELPFUL HINTS

→ Run this lab in a space that is safe and large enough for plenty of running (grassy terrain in a large backyard or park will work best).

PROCEDURE

1. Stretch your 100-meter tape measure out along a flat grassy area, pulling it tight on both ends (you may want to anchor it at both ends) **(Fig. 1)**.

2. Use your timer to record the time it takes you and your friend(s) to run the 100-meter stretch (one runner at a time). Runners can try the course more than once in order to obtain their best time. Record each runner's best time in your lab notebook **(Fig. 2)**.

3. Because a cheetah's prey rarely runs in a straight line as it tries to escape, cheetahs often have to turn sharply and zigzag across the landscape during an active hunt. With this in mind, set some obstacles out along the 100-meter path, similar to a slalom course, and time your runners again. This time they must bob and weave through the obstacles as they make their way to the finish line. Again, record each runner's best time in your lab notebook **(Fig. 3)**.

4. Consider calculating your running speed so you can compare it to other animals. To do this, you'll need to use the formula speed=distance/time. For example, if you ran 100 meters in 30 seconds, your running speed would be 3.3 meters per second (100/30=3.3). Most running speeds are recorded in terms of miles per hour, so you'll need to convert your distance to miles (meters run/1,609) and your time from seconds to hours (time run/3,600). If you ran 100 meters in 30 seconds, that would equal 7.5 miles (12 km) per hour. How do you measure up?

5. Consider turning your data into colorful graphs or charts to share.

Fig. 1

Fig. 2

Fig. 3

THE
SCIENCE
BEHIND THE FUN

→ Predatory species across the animal kingdom use a wide range of hunting strategies to find and catch their prey. Cheetahs, often called pursuit predators, use speed and agility to outmaneuver their prey. Long legs, a flexible spine, and partially unretractable claws help them grip the surface beneath them. Amazingly, cheetahs can reach speeds of 75 miles per hour (121 km), with the fastest-recorded 100-meter cheetah sprint clocked at 5.95 seconds. In comparison, 9.63 seconds is the fastest recorded 100-meter sprint by a human (Jamaica's Usain Bolt in 2012). In this lab, you'll discover how the sprinting abilities of you and your friends measure up.

CREATIVE ENRICHMENT

→ While cheetahs quickly reach unmatchable speeds over short distances, they are not especially well-built for long-distance running. Other species excel at maintaining their pace over long distances, often as a means to outlast and exhaust prey. These persistence hunters include a variety of canid species, including African wild dogs and gray wolves. Consider designing a longer course and see how well you and your friends can keep pace.

SWIM STUDY

Take a closer look at how fins make for successful swimming.

→ Consider using a digital camera or other handheld device to capture video of the fish you're observing so you can go back and review its movement in slow motion.

PROCEDURE

1. Choose an individual fish to observe. If multiple fish are present, select an individual with unique markings or coloration so you don't lose track of it.

2. Sketch your fish in your lab notebook, labeling all fins and other body structures. Carefully observe each fin or fin pair, recording the way it moves in the water. Does it move up and down like wings? Does it move forward and backward? Does it move side to side **(Fig. 1)**?

3. Introduce a few fish flakes to the top of the aquarium or fish bowl and watch carefully to see how the fish uses its fins to move to the surface. Record your thoughts about how the fins moved and which fins were important for this type of motion **(Fig. 2)**.

MATERIALS

→ Access to a fish in an aquarium or clear fish bowl
→ Fish flakes
→ Fish gravel

SAFETY TIPS & HELPFUL HINTS

→ If you don't have a fish of your own, consider visiting someone's house where there are fish present, or maybe save up to buy one or two of your own!

Fig. 1

Fig. 2

4. Gently drop a piece of aquarium gravel into the water. Watch carefully to see how the fish uses its fins to move out of the way of the incoming object and then investigate it once it lands on the bottom. Record your thoughts about how the fins moved and which fins were important for this type of motion.

CREATIVE ENRICHMENT

→ As mentioned earlier, fish come in all different shapes and sizes, and many fish have modified fin structures different from the ones you've just observed. Consider taking a trip to your local zoo or aquarium to observe fishes in a larger setting with diverse species, especially those with unique fin structures, such as rays and skates, eels, flatfish, and many reef species, such as angelfish and butterflyfish. How do these specialized species make their way in the water?

THE SCIENCE BEHIND THE FUN

→ While cheetahs are fastest on land, there are fish in the ocean that can swim even faster (black marlins can swim 80 miles [129 km] per hour!). Have you ever paid close attention to the movements of a fish in an aquarium? Swimming as a primary mode of locomotion is actually quite complex, due in large part to the difference in density between air and water, so fish have evolved several structures to make continuous swimming possible.

While the shapes and sizes of fish bodies vary, nearly all have fins that allow them to reposition the water around their bodies strategically. Each fin plays a particular role in the fish's underwater movement. The dorsal fin helps it stay balanced and upright. The pectoral fins help direct the fish's movement up and down and side to side. The pelvic fins help the animal move up and down in the water, as well as make sharp turns and quick stops. The anal fin helps the fish stay stable. The caudal fin (or tail fin) is the only fin directly connected to its spinal column. It's what gives the fish its speed, propelling it forward. In this lab, you'll take a closer look at how fish use their fins to displace water around them and successfully navigate.

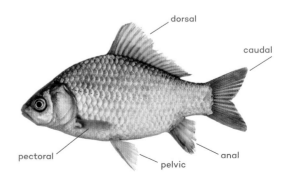

LAB 32

SWIMMER'S CHALLENGE

Investigate how appendages with increased surface area support faster swimming.

MATERIALS

→ Swim fins
→ Pool (or other swimmable body of water, such as a pond or lake)
→ Timer
→ Assistant

SAFETY TIPS & HELPFUL HINTS

→ This lab requires adult supervision.

→ You should only try this lab if you're a confident swimmer and you know how to swim using fins.

PROCEDURE

1. Stand in water deep enough to fully submerge your arms. Under the water, hold both of your hands horizontally a few inches away from your chest and about 6 inches (15 cm) below the surface, flat palms facing the bottom of the pool. Move your hands together and then apart several times to see what happens at the surface of the water **(Fig. 1)**.

2. Turn your hands to a vertical position and try the same motion again at the same speed to see if there is a difference at the surface of the water. The vertical position creates a greater surface area and should therefore displace more water. Record your thoughts and observations in your lab notebook **(Fig. 2)**.

3. Have your assistant time you as you swim as fast as you can from one side of the pool to the other.

4. Take a few minutes to let your body reset.

5. Put on your swim fins and repeat step 3. Record both times in your lab notebook. You may consider doing both swim trials multiple times to record your best time. How did the use of fins affect your swim speed **(Fig. 3)**?

6. Consider calculating your swim speed so you can compare it to that of other swimming creatures. To do this, you'll need to measure the length of the pool and use the formula speed=distance/time. For example, if you swam 20 feet (6.1 m) in 10 seconds, your swim speed would be 2 feet (0.6 m) per second (20/10=2). But most swim speeds are recorded in miles per hour, so you'll need to convert your swim distance from feet to miles (feet swum/5,280) and your swim time from seconds to hours (time swum/3,600). So, if you swam 20 feet (6.1 m) in 10 seconds, that would equal 1.3 miles (2.1 km) per hour. How do you measure up **(Fig. 4)**?

7. Consider turning your data into colorful graphs or charts to share.

Fig. 1

Fig. 2

Fig. 3

Fig. 4

THE
SCIENCE
BEHIND THE FUN

→ While fish represent the largest group of vertebrates on the planet, with more than 30,000 described species, they are certainly not the only animals that live in water. Lots of animals make their homes in aquatic environments, some for their whole lives (fully aquatic) and others only during certain periods of life (semi-aquatic). These include amphibians, birds, reptiles, and mammals, not to mention an enormous number of aquatic invertebrates.

An aquatic lifestyle often requires special body features, such as gills for breathing or modified tooth structures for filter feeding. For some species, fish fins don't cut it. Many use modified appendages to maximize their chances of survival in aquatic habitats. Animals such as whales, sea turtles, seals, and sea lions have flippers, while some amphibians, birds, reptiles, and mammals rely on webbed feet. Like your hands when you held them vertically in the water, the tails of crocodilians (crocodiles, alligators, and their relatives) provide increased surface area so each movement displaces more water, moving the animal faster with less effort. These adaptations are critical for capturing prey and evading predators. In this lab, you'll investigate how using artificial fins effects swimming speed.

CREATIVE ENRICHMENT

→ Visit your school library or do some online research to learn more about the variety of incredible features that make an aquatic lifestyle work. Good study candidates include dragonflies, diving beetles, sharks, alligators, turtles, sea snakes, frogs, penguins, ducks, whales, and beavers.

THE WONDERS OF WINGS

Discover how wings help flying animals take to the air.

MATERIALS

→ Large piece of cardboard or poster board
→ Scissors
→ Measuring tape
→ Masking tape or painter's tape
→ Several pieces of lightweight material (cotton balls, Styrofoam packing peanuts, shreds of paper, etc.)

SAFETY TIPS & HELPFUL HINTS

→ Be careful when choosing the tape to affix the "wings" to your arms. Some will be easier to remove from your skin than others (do *not* use duct tape! Ouch!). You might also consider wearing snug-fitting long sleeves for this activity.

PROCEDURE

1. Outline a 3-foot x 3-foot (0.9 m x 0.9 m) square on a flat floor with tape **(Fig. 1)**. Spread several small pieces of lightweight material everywhere in the square except the very middle **(Fig. 2)**.

2. Stand in the middle of the littered square. Stretch out your arms to each side and flap them vigorously like a bird 10 times (count out loud). Carefully record in your lab notebook the number of pieces that blew outside the square while you flapped. Put all material back inside the square to reset the grid **(Fig. 3)**.

3. Cut the cardboard or poster board into two wing shapes the length of each of your arms and tape them in place.

4. Stretch out your "wings" and flap them vigorously like a bird 10 times (count out loud). Carefully record in your lab notebook the number of pieces that blew outside the square while you flapped **(Fig. 4)**.

5. Consider turning your data into colorful graphs or charts to share.

Fig. 1

Fig. 2

Fig. 3

Fig. 4

CREATIVE ENRICHMENT

→ After seeing firsthand the amazing force that wings exert on the air, it is also fascinating to discover how *rapidly* flying animals can flap their wings. Hummingbird wings can beat an astounding 50 times per second, flies can achieve more than 200 wing beats per second, and some bees can complete more than 500 beats in a single second! See how many times *you* can flap your own wings for comparison.

THE SCIENCE BEHIND THE FUN

→ Animals with an aerial lifestyle are generally split into two categories: powered or active flight (using energy) and unpowered or passive flight (using no energy). Only three living animal groups use powered flight: insects, birds, and bats. Flight in these groups evolved at different times in Earth's history, and the wing structures vary in interesting ways. Depending on the species, most insects have one or two pairs of wings, which are actually part of their exoskeleton and contain no bones. In contrast, the bones in a bird's wing are its arm bones—humerus, radius, ulna—while those of a bat are the hand and finger bones—metacarpals and phalanges.

Another difference is that bird wings are covered in feathers, which are incredibly dynamic structures themselves, while bat wings are covered by a thin sheath of skin that connects the bones to each other and to the body. Insects, birds, and bats all use their unique wing structures to navigate important forces, such as lift, thrust, and drag, in order to take flight and maximize their use of aerial resources. Just as the displacement of water by fins is key to swimming and floating, so too is the displacement of air by wings, especially during takeoff. In this lab, you'll explore how wings help lift animals into the air.

BUILD YOUR OWN FLYING SQUIRREL

Explore the strategy of gliding by building your own flying squirrel.

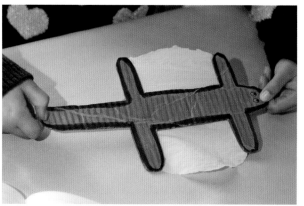

Fig. 4

MATERIALS

→ Cardboard
→ Scissors
→ Markers and crayons (optional)
→ Clear tape (as lightweight as possible)
→ Coin or other small weight
→ Piece of paper or plastic bag
→ Timer

SAFETY TIPS & HELPFUL HINTS

→ You might need to try a few different coins to see which works best as a weight for your squirrel, but be sure you use the same weight for each gliding trial (with and without patagium present).

→ Consider drawing a diagram of your flying squirrel in your lab notebook before you start building. Be sure to label all body parts and intended materials (Fig. 1).

→ Ask permission to use YouTube for the Creative Enrichment section of this lab, and be sure that Restricted Mode is enabled under Settings to avoid any inappropriate content.

Fig. 1

PROCEDURE

1. Draw the shape of a small squirrel on your cardboard, including its head, body, front and hind limbs, and tail. Feel free to draw a face and decorate your squirrel with crayons or markers, but try not to add weight.

2. Cut out your masterpiece.

3. Tape the metal washer to the center of the underside of the squirrel's belly to establish a center of gravity and keep it upright during flight (Fig. 2).

4. From a high point such as a balcony or the top of a ladder (be careful!), drop the squirrel and use your timer to record how long it takes to hit the ground (Fig. 3).

5. Cut out two patagia from your lightweight paper or plastic and tape one to each side of the squirrel. A patagium (patagia is the plural) is the term for the skin that stretches from a flying squirrel's ankles to its wrists and allows it to glide hundreds of feet.

6. Drop your squirrel from the same place you used before, and use your timer to record its new flight time (Fig. 4).

7. Consider turning your data into colorful graphs or charts to share.

Fig. 2

Fig. 3

CREATIVE ENRICHMENT

→ Scientists are learning more about the flight strategies and associated body features of flying animals all the time. Many of these unique creatures can be seen demonstrating their skills in nature videos, and are simply amazing to watch. With your parents' permission, log onto YouTube and search for flying squirrel, flying dragon lizard, flying frog, sugar glider flight, parachuting spiders, flying snake, and flying fish. Record your thoughts about each video clip and share what you learn.

THE
SCIENCE
BEHIND THE FUN

→ Not all flying animals are built for powered flight. Some take advantage instead of aerodynamic body shapes and special features that allow them to move through the air without exerting much energy. The three main strategies for unpowered flight in animals are soaring, parachuting, and gliding.

Many bird species save energy by soaring on pockets of rising air to gain or maintain the altitude reached through powered flight. A variety of spiders drift in the wind on miniature silk parachutes, and some frog species parachute using the webbing on their feet.

Gliding animals launch themselves into the air and then use specialized skin flaps or body movements to glide over short distances. Flying fish, flying frogs, flying lizards, flying snakes, flying squirrels, and the gliders of Australia and New Guinea all use this strategy. Flying squirrels, found in North and South America and parts of Europe, launch themselves from trees, patagia outstretched, and use their limbs and fluffy tails to steer their bodies. In this lab, you'll build your own flying squirrel and investigate the importance of its body features for gliding.

INTERESTING ANIMAL
FAMILIES

Species live in families and other social groups in order to work together to achieve common goals, such as escaping predators, finding food, conserving energy, and raising young.

Meerkats and other species use sentinels to warn the group about danger, impala herds scatter to confuse their attackers, and gulls work as a team to drive away predators. Ants use chemical messages to report food resources to the group, marlins work together to herd prey into a common feeding area, and wolves work cooperatively to take down prey. Geese and other large flocking birds fly in aerodynamic formation to minimize drag, sardines swim in hydrodynamic formation to minimize drag, and penguins huddle together in large groups to conserve and share collective heat. And, to maximize their chances of survival, offspring of elephants and other species receive communal care, regardless of who the parents are.

The labs in this unit will help you discover how animals rely on family structures as they produce and protect offspring. From researching and constructing the most durable nest to calculating the impact of increased litter size, this unit gives you the chance to appreciate the advantages and sacrifices associated with family life.

SUCCESS IN NUMBERS

Investigate the benefit of group formation for conserving body heat.

MATERIALS

→ Group of friends or family members
→ Timer

SAFETY TIPS & HELPFUL HINTS

→ Try this lab on a cold evening, but be sure not to put yourself or others in any danger by staying out too long.

PROCEDURE

1. On a clear and chilly evening after dark, invite your friends or family members to join you outside for a few minutes (no jackets—that's cheating!). Ask the group members to stand at least 5 feet (1.5 m) apart for 3 minutes, then ask them for feedback on how exposure to the elements affected their bodies. Record their thoughts and your own in your lab notebook **(Fig. 1)**.

2. Find the coldest person in the group and put them in the middle of a huddle (group hug). Stay in this formation for another 3 minutes, then record each person's feedback on how the exposure affected their bodies this time **(Fig. 2)**.

3. Have group members take turns cycling into the center of the huddle for 3 minutes. Again, record each person's feedback on the effects of exposure.

CREATIVE ENRICHMENT

→ Besides their group huddling skills, emperor penguins are some of the most dedicated parents on the planet, making incredible sacrifices for their chicks. They commit themselves to tireless hours of egg protection and incubation in freezing cold habitats and travel enormous distances in predator-infested waters to find and bring back food. Consider taking a few minutes to write a thank you note to your parents or guardian for the important contributions they've made to *your* life so far.

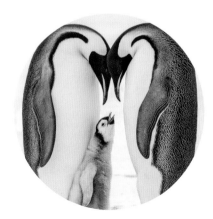

Fig. 1

Fig. 2

THE
SCIENCE
BEHIND THE FUN

→ Living in a group can have many advantages, including sharing warmth. Emperor penguins survive winter in the coldest parts of the world where there is very little shelter from the wind and cold. They can sustain some of the largest animal colonies on the planet because of their willingness to huddle together. The huddle formation allows each individual to conserve body heat and use less energy to stay warm. The huddle cycles each penguin from the outside to the inside so no one is left in the cold for long. In this lab, you'll explore how working together with others can improve your chances of staying warm in a cold environment.

THE PERFECT NEST

Build your own nest to see how well it stands up to the elements.

MATERIALS

→ Variety of natural objects (sticks, twigs, pine needles, leaves, dried grasses, rocks, feathers, mud, etc.)
→ Newspaper
→ Paper plate
→ Empty plastic eggs
→ Hair dryer
→ Watering can

SAFETY TIPS & HELPFUL HINTS

→ Use only nest materials that are readily available to wild birds (no tape or glue—that's cheating!). However, you may use non-natural items, such as string, pieces of cloth, and shredded paper, because birds often make use of these **(Fig. 1)**.

Fig. 1

Fig. 2

PROCEDURE

1. Lay out your items on a flat work surface covered with newspaper or a tablecloth. Think about what type of nest you want to build, and make a rough sketch in your lab notebook. As you perfect your design, ask yourself: Will eggs roll out of this nest? Will eggs fall through this nest? Will this nest fall apart under strong winds? Will this nest fill with water in the rain **(Fig. 2)**?

2. On top of a paper plate, build your nest from the materials you gathered, trying different arrangements and designs until you feel confident that your nest will support a clutch of eggs (or a brood of chicks).

3. Place a few plastic eggs in your nest and see if they stay inside (don't roll out).

4. Gently lift your nest off the paper plate and see if the eggs fall through in any place.

continued on page 98

CREATIVE ENRICHMENT

→ Now that you've learned a bit about the most common nest types, consider going on a nest scavenger hunt in your neighborhood or at a local park. Remember that nests can be found on the ground, inside bushes, in tree trunks, and high up in the branches, so be sure to look in lots of different places. Whatever you do—*do not touch* the nests you find. Nests tend to be quite fragile and can contain harmful bacteria, so it's best to take a photo or make a sketch of any nests you find, rather than handling them directly. Share your findings!

Fig. 3

5. While holding the nest above the paper plate, use a hair dryer on its highest setting to see if the eggs will blow out of the nest on a particularly windy day (blow from the top, sides, and bottom to test this) **(Fig. 3)**.

6. Use the watering can (do this outside!) to gently pour water over your nest to see if it has proper drainage **(Fig. 4)**.

7. Consider trying different nest designs to see which stands up best to these artificial weather elements. Be sure to sketch each design in your lab notebook and record how well each design performed.

Fig. 4

THE
SCIENCE
BEHIND THE FUN

→ While emperor penguins make incredible sacrifices for their chicks, one thing they do *not* do is build a nest. Instead, they balance their one precious egg on the top of their feet while keeping it warm against a special flap of skin called a brood patch. This fact makes emperor penguins quite different from nearly all other birds, as well as various invertebrates, fishes, amphibians, reptiles, and mammals, which build several different nest types in order to protect their eggs and young from predators and environmental factors.

Nest designs vary from simple to complex, and are made using lots of different materials. Falcons, shorebirds, and ostriches use simple scrape nests—a small, uncovered, depression on the ground or in plants **(Fig. A)**. Owls, parrots, and woodpeckers use cavity nests—a hole in the trunk of a tree or cactus **(Fig. B)**. Puffins, kingfishers, and petrels use burrow nests consisting of a tunnel and small chamber dug into cliffs and hillsides **(Fig. C)**. Flamingoes build mound nests of mud and decomposing plant material that helps keep eggs warm **(Fig. D)**. Ospreys, herons, and egrets build large, flat platform nests in treetops or on the ground **(Fig. E)**. Orioles and weavers build sac-like pendant nests made from grasses and twigs that hang from branches **(Fig. F)**. And swallows, hummingbirds, and warblers build common cup nests, which look like small bowls and are usually made with twigs and dried grass held together by mud or saliva **(Fig. G)**. In this lab, you'll design and build your own bird nest and see how well it holds up against simulated elements.

Fig. A

Fig. B

Fig. C

Fig. D

Fig. E

Fig. F

Fig. G

THE MAGNIFICENT EGG

Take a closer look at the shape, structure, and strength of bird eggs.

MATERIALS

→ Chicken eggs (from a carton)
→ Plastic tablecloth or other protective covering
→ High-powered flashlight
→ Books
→ Push pin
→ Scalpel or small scissors
→ Clear tape

SAFETY TIPS & HELPFUL HINTS

→ You might need to ask for help from a parent or older sibling to cut your eggshells. Don't be discouraged if it takes a few tries!

PROCEDURE

1. Examine the outside shell of an egg with your magnifying glass. Can you see the tiny pores? Sketch the egg in your lab notebook. What shape is it?

2. Hold the egg at eye level with one hand and shine a high-powered flashlight through the egg to examine its insides. You should be able to see the yolk (provides food for the chick), the white (provides a cushion for the chick as it grows), and the air pocket (provides a place for exchange of

Fig. 1

oxygen and carbon dioxide). Sketch the lighted egg in your lab notebook and label the parts **(Fig. 1)**.

3. Hold the egg between your hands with the pointed ends against your palms. Make sure you do this over the plastic tablecloth! Press your hands together as hard as you can **(Fig. 2)**.

4. Place two eggs next to each other on the tablecloth, about 2 inches (5 cm) apart. Cover them with a small layer of the tablecloth. Carefully stack a book on top of the two eggs. Keep stacking books gently until the eggs break. Record the number and sizes of the books that the two eggs held **(Fig. 3)**.

5. Take a fresh egg and carefully poke a hole in each of the narrow ends with a push pin. Working over a bowl, blow into one end of the egg so that its contents flow out.

Fig. 2

Fig. 3

Fig. 4

Fig. 5

6. Put a piece of clear tape around the center of the egg and carefully cut in a complete circle, so you end up with two domes. Gently remove the tape and clean inside and outside the shells **(Fig. 4)**.

7. Repeat this process with a second egg so you end up with four dome structures. Try to make each dome the same height.

8. Place your four egg domes so their open centers are against the tablecloth in a 4 inch by 6 inch (10 cm x 15 cm) rectangle. Carefully stack books one at a time onto the egg domes until one of them breaks. Record the number and sizes of the books that the four egg domes held **(Fig. 5)**.

CREATIVE ENRICHMENT

→ Bird eggs vary greatly in size, from the bee hummingbird egg (about the size of a jelly bean) to the ostrich egg (6 inches [15 cm] tall and up to 5 pounds [2.3 kg]). They also come in a rainbow of colors and speckle patterns. Visit the library or do some online research to learn more about the diversity of bird eggs.

THE SCIENCE BEHIND THE FUN

→ If you've ever been lucky enough to come across a nest with eggs in it, you know how exciting that is—just to think that tiny chicks are developing safely inside. Bird eggs are different from most other animal eggs in that their shells are hard and durable. In fact, that dome shape happens to be one of the strongest structures in the world, as it distributes weight and pressure applied to the top evenly throughout the entire structure. This is due in large part to the crystalline structure of calcium carbonate, their primary building material.

Durability is good since most bird parents sit on top of their eggs to incubate them. But chicks must be able to hatch out of the egg on their own, so the shell also can't be too strong. The egg serves not only to protect the chick, but also to nourish it as it grows and develops and allows it to breathe through thousands of tiny pores distributed across the shell. In this lab, you'll take a closer look at common chicken eggs and discover just how strong they can be.

ALLIGATOR INCUBATOR

Explore the benefits decaying plants offer alligator hatchlings.

MATERIALS

→ Digital meat thermometer
→ Plant parts (stems, branches, leaves, bark, etc.)
→ Dirt or soil
→ Watering can
→ Bright cloth or flag

SAFETY TIPS & HELPFUL HINTS

→ This lab takes a few weeks to complete, so be patient and remember to take your temperature readings every week.

→ It's best to run this lab in the middle of summer so that your daily temperatures remain relatively consistent. This is also when alligators build their nests in the wild.

→ When choosing your nest location, make sure you clear it with whoever maintains your yard!

PROCEDURE

1. Build a mound of dead plant material in a safe, sunny spot in your yard. Start with a 1 inch (2.5 cm) layer of plant material, add a thinner layer of soil or dirt, and then alternate between the two until you build up five or six layers of each type and have a good-sized mound **(Fig. 1)**.

2. Water the mound.

3. Stick the metal probe of the meat thermometer into the middle of the mound, as deep as it will go. Mark this temperature recording spot on the mound with a brightly colored cloth or flag so you always take the temperature from the same spot. Record the starting temperature in your lab notebook **(Fig. 2)**.

4. Measure the temperature at your marked spot twice a week for four weeks. After each recording, pour a fresh watering can of water on the mound to help continue the decomposition process.

5. Consider turning your data into colorful graphs or charts to share.

Fig. 1

Fig. 2

THE
SCIENCE
BEHIND THE FUN

→ Birds usually come to mind first when we think about egg layers, but actually, species all across the animal kingdom produce them. But not all animals incubate their eggs in the same way that birds do.

Nearly all bird species lay their eggs in a nest and then incubate them using their own body heat, but many animals simply lay their eggs and say goodbye. Other species incubate their eggs *inside* their bodies; hatching occurs internally.

Still others, such as alligators, lay their eggs in a nest but, instead of relying on their own body heat for incubation, use outside elements to keep the eggs at an optimal temperature. Female alligators build mound nests out of plants, sticks, leaves, and mud near the water's edge before laying their round, white eggs. They then cover their eggs with layers of dead plant material over the roughly two-month incubation period. As tiny microorganisms work to break down the layers of plant material, they produce heat; incredibly, this heat determines the gender of the hatchlings inside the nest! Warmer nests produce more males, cooler nests produce more females. In this lab, you'll document the heat produced by plant decomposition in a simulated alligator nest.

CREATIVE ENRICHMENT

→ Not all animals build nests for their eggs ... some build them for themselves! Gorillas, for example, use leaves and branches to build a new nest on the ground or in the trees every evening for comfort, warmth, and protection. Consider building a gorilla nest in your backyard to see how comfortable, warm, and hidden you can make it.

BABY ON BOARD

See how hard it can be to bring babies along for the ride.

MATERIALS

→ Young partner (ideally a toddler, two to three years old)
→ Cloth bag or tote large enough to hold your partner and with a handle that can go around your waist

SAFETY TIPS & HELPFUL HINTS

→ While you experiment with toting babies around in this lab, be sure not to hurt yourself (or your young partner) in the process. Know your limits and take breaks when you need them!

→ Be careful not to choke your partner in step 3!

→ Before using Google, enable the "SafeSearch" feature to block any inappropriate images.

PROCEDURE

1. Get down on the floor on your hands and knees. Ask your partner (or two) to climb onto your back and hold on tight, just like a baby gorilla would do. Now try making your way around the house for at least three minutes: up and down the stairs, onto the couch, up onto your bed, into the kitchen to grab a snack from the pantry, and on different surface types. Record in your lab notebook your thoughts on what was easiest and most challenging about the process. What parts of your body was this most hard on? Which tasks were simply impossible to accomplish? Also get feedback from your passenger(s) **(Fig. 1)**.

2. Get down on the floor on your hands and knees again, but this time have your partner(s) cling to your front so they are hanging on like a baby sloth. Record your thoughts on the process and get feedback from your partner **(Fig. 2)**.

3. Get down on the floor on your hands and knees again, but this time grab the back of your partner's shirt with your mouth and try carrying (or dragging) them around the house. Record your thoughts on the process and get feedback from your partner **(Fig. 3)**.

4. Put one handle of the large cloth bag around your waist and have your young partner climb into the bag. Hold the other bag handle tightly as you squat down on your legs like a kangaroo. Try hopping around the house with this added load. Record your thoughts on the process and get feedback from your partner **(Fig. 4)**.

Fig. 1

Fig. 2

Fig. 3

Fig. 4

THE
SCIENCE
BEHIND THE FUN

→ Whether hatched from an egg or born live, animal newborns are not always self-reliant right away. Many species care for their babies for a period of time, sometimes years. Animal parents and other caring group members help protect babies from danger, provide food and nourishment, and teach the young how to survive in a complex world. While some babies must be ready to walk and even run within minutes to ensure survival (zebras, antelope, elephants, etc.), many species have efficient ways to bring their babies around with them. Quite a few species carry their babies on their backs, both on land and in the water (apes, opossums, ducks, loons, geese, insects, spiders, etc.). The babies of other species cling onto their parent or babysitter so their caregiver can use all four limbs to climb and move around (monkeys, sloths, etc.). Some species carry their babies using their mouths (dogs, cats, alligators, etc.), and some have evolved special pouches to hold their babies (kangaroos and other marsupials). In this lab, you'll try carrying around young members of your group to keep them safe and sound.

CREATIVE ENRICHMENT

→ Carrying one or two young partners around was hard work, but imagine carrying 10, 50, or even 100 babies around with you! Search Google images for opossum carrying babies, scorpion carrying babies, whip scorpion carrying babies, and spider carrying babies. It's pretty incredible to see just how dedicated animal parents can be!

LAB 40

MORE MOUTHS TO FEED

Discover how much food it takes to support a large family.

PROCEDURE

1. Record in your lab notebook your daily food intake for one full week—write down everything you eat each day for seven days—starting on a Monday. Label the days so you can compare them.

2. Calculate your daily average food intake for the seven-day period. For example, if you ate five crackers on Monday, seven on Tuesday, five on Wednesday, four on Thursday, six on Friday, nine on Saturday, and eight on Sunday, you would add them all (5+7+5+4+6+9+8=42) and divide by seven days (42/7=6) to discover that your average daily intake of crackers is six **(Fig. 1)**.

3. Now imagine that you're a naked mole rat with 30 offspring in one family. How much food would you need to support everyone for a single day?

4. Consider calculating the overall food cost in addition to overall amount for 30 siblings. For example, if the box of crackers you've been eating all week costs $6.00 and there are 100 crackers in the box, then each cracker costs six cents ($6.00/100=$0.06).

MATERIALS

→ Calculator
→ Lab notebook
→ Pen or pencil

SAFETY TIPS & HELPFUL HINTS

→ Some foods are easier to record in whole numbers, and others are easier to record as number of cups or other units. For example, record number of crackers eaten but use cups or ounces for pasta or cheese. Also, it's fine to approximate the amount, giving it your best estimate.

Fig. 1

THE
SCIENCE
BEHIND THE FUN

→ Just like parental care varies across species, so too does litter size. Elephants birth only one baby every few years, whereas mice can produce up to 100 babies through multiple birth cycles in a single year. And that's just for mammals! Amphibians, fishes, and invertebrates can produce hundreds, thousands, and even millions of offspring in a year.

Multiple factors influence litter size, including mode of reproduction, offspring size, resource availability, and predation risk due to other species. To successfully pass their genes on to future generations, many animals produce large litters—naked mole rats can have up to 30 babies in a single litter! But at what cost? While animals don't have to worry about clothing, daycare, school supplies, and toys for their babies, the universal truth is that more babies mean more mouths to feed. In this lab, you'll explore the cost of increasing litter size.

CREATIVE ENRICHMENT

→ Many species live in regions of the world where access to fresh water is limited. Consider using the same procedure above to calculate your average daily water intake. How much water would it take to support a 30-offspring family for a single day?

LIVING ALONGSIDE
ANIMALS

While most people can recognize a polar bear or orangutan, very few will ever see one in the wild. Some of the most fascinating creatures in the world are not those found on the other side of the planet but the ones found in your own backyard. Some of life's most memorable moments come from simple interactions with the animals that live right around us.

It's easy to go outside on any given day and find animals living in your neighborhood. Whether you're birdwatching with binoculars, counting bees on flowers, looking at lizards, or chasing squirrels up trees, familiar species are out and about while the sun is up (diurnal). But many of the animals that live around our homes are only active at night (nocturnal), so it takes special preparation and patience to learn more about their lives.

The labs in this unit will help you seek out and interact with the animals that live near your home. From raising and releasing butterflies to conversing with frogs under the stars, this unit creates opportunities for up close and personal interactions with your neighborhood species.

NIGHTTIME ANIMAL SEARCH

Head outside to document the sights and sounds of nocturnal animals.

MATERIALS

→ Flashlight
→ Digital camera or handheld device with camera (smartphone, tablet, etc.)
→ Partner (optional)
→ Insect repellant (optional)

SAFETY TIPS & HELPFUL HINTS

→ Do this lab with at least one partner so you're not heading out in the dark all alone. But be sure to choose someone quiet so they don't scare away all the animals!

→ You're likely to see and hear more animals if you do this lab during the spring, summer, or fall months. Once winter sets in, many local animals will hunker down to wait out the cold.

PROCEDURE

1. Head outside with your flashlight (and insect repellent, if you need it) at least one hour past sunset to be sure it's completely dark out. Find a quiet, comfortable spot in your yard and sit down in grass or soft dirt. Close your eyes and listen to the sounds around you. Try to tune out car engines, sirens, neighbors playing music, and the like so you can hear the sounds of nature. What animal sounds do you hear? Can you hear crickets or other insects? Owls? Birds? Bats? Coyotes? Wolves? Frogs? Use your flashlight to see so you can record your thoughts in your lab notebook. How many different animal sounds were present? How many different human-made sounds did you hear? Also be sure to record the time of night, the time of year, the size and position of the moon, and weather conditions such as clouds and wind.

2. Point your flashlight in the direction of an animal sound you hear and see if you can spot its source from where you're sitting **(Fig. 1)**.

3. Walk around your yard with your flashlight to search for animals and their homes. Look up in the trees, on tree bark, inside of bushes, under rocks, on different ground surfaces, on the sides of your house, up in the eaves, and near your outdoor lights where insects and other invertebrates like to gather. Record all animal sightings in your lab notebook and consider taking photos of the animals that you see, whenever possible **(Fig. 2)**.

Fig. 1

Fig. 2

THE
SCIENCE
BEHIND THE FUN

→ Nocturnal animals have plenty of good reasons for making their way at night: they face less competition for resources, they save energy and water by being active in cooler temperatures, and the darkness provides a form of camouflage from many predators. Who in your neighborhood uses this strategy to their advantage? In this lab, you'll head outside at night to discover and document nocturnal animals around your home.

CREATIVE ENRICHMENT

→ Fireflies are one of the most common (and exciting) nighttime animals to observe. There are more than 2,000 species of these magical little beetles distributed throughout the world and across many different habitats. Fireflies produce light through bioluminescence to attract mates and prey, displaying in shades of green, yellow, pink, and orange. When fireflies are active, you can often get a response from them by producing a series of artificial flashes. Sit quietly in the grass at night and use a small flashlight or pen light to produce quick flashes in succession with pauses in between. If you get a response, see how long you can keep the interaction going. It's also fun to capture fireflies gently in a clear jar so you can make closeup observations. Be sure to poke holes in the lid for air and don't keep them confined in the jar for more than a few minutes.

CALLING ALL OWLS

Set out on an owl hunt to discover the habits of local species.

MATERIALS

→ Flashlight
→ Recorded owl calls available online and in apps
→ Handheld device capable of playing owl call recordings in speaker mode
→ Binoculars
→ Gloves
→ Tweezers
→ Handheld magnifying glass
→ Partner (optional)

SAFETY TIPS & HELPFUL HINTS

→ Do this lab with a partner so you're not out in the dark alone. But be sure to choose a quiet partner so they don't scare away all the owls!

→ Some species are active right after sunset while others wait until it's completely dark out, so you'll need to commit a few hours to the search.

→ If you can't find any owl pellets in your neighborhood, you can order them from online retailers.

PROCEDURE

1. Head outside just after sunset and look up into as many trees as you can to spot resident owls. Also look at the base of each large tree you come across to see if you can find any owl pellets (they look like little packets of bones and fur). If you find any, collect them with a plastic bag (use it as a glove so you don't touch the pellets) and record in your lab notebook where you found them **(Fig. 1)**.

2. Stop occasionally and listen for owl sounds. If none are present, play your owl sounds on speaker. You can also play these sounds once you return

home to see if you can coax an owl into your yard. Find a comfortable spot to sit where you can see lots of trees and then play the recordings one at a time **(Fig. 2)**.

3. If you locate an owl on your hunt or in your backyard, use your flashlight and binoculars so you can make a rough sketch of it in your lab notebook. Note its size, coloration and markings, beak shape, and what it's doing. If you can identify what kind of owl it is, try playing back sounds of that species to see if it reacts in some way. If you can, get closer to the owl to capture more details with your flashlight and binoculars, until the owl flies away **(Fig. 3)**.

4. Once you're back inside, put on protective gloves and carefully dissect the owl pellets with your tweezers, gluing the bones onto a piece of white paper. Use your handheld magnifying glass to carefully examine each bone fragment and try to identify the area of the body that it came from (legs, arms, vertebrae, skull, teeth, etc.). Can you identify the type of animal that this owl ate **(Fig. 4)**?

Fig. 1

Fig. 2

Fig. 3

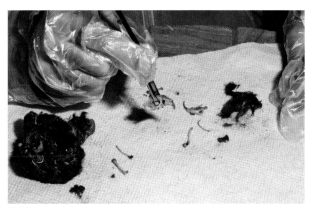
Fig. 4

THE SCIENCE BEHIND THE FUN

→ With more than 200 species worldwide, owls are a common sight in many neighborhoods. Perfectly adapted for nighttime hunting, owls easily ambush their prey thanks to special serrations on their feathers that make them almost completely silent while flying. Their ears are in different positions on each side of the head, allowing them to triangulate precisely the sounds of prey movement. Their eyes are so large they can't turn in their sockets, so the owl has to turn its head to see around. But those eyes offer impressive distance and low-light vision, allowing owls to spot potential prey from afar. Most owl species are so well camouflaged they seem to disappear into their habitat, and their extremely sharp claws and beaks capture and rip apart their prey. Owls are basically stealth hunting machines! In this lab, you'll set out on an owl hunt and see if you can learn more about the diet of local species.

CREATIVE ENRICHMENT

→ Interestingly, not all owls are strictly nocturnal. For example, burrowing owls are active in the day *and* the night. They often make use of burrows dug by other species, such as prairie dogs and ground squirrels, for roosting and raising young. Burrowing owls range across North and South America, but you can see them instantly by visiting one of many burrowing owl cams online. Check them out!

PICKY BIRD BEAKS

Document the food preferences of wild birds around your home.

MATERIALS

→ Raised platform, balcony railing, or standing bird bath (dry)

→ Variety of safe and appropriate wild bird foods (unshelled unsalted sunflower seeds, live or dried meal worms, bird seed, freshly cut melon, halved grapes, unsalted peanuts, small cup of sugar water, etc.)

→ Binoculars

→ Bird guide or other bird identification resources (optional)

SAFETY TIPS & HELPFUL HINTS

→ Choose a raised platform you and the birds can access easily, and make sure you can see it from a comfortable spot nearby. You can use an existing bird bath as long as it's dry or a balcony railing if it is wide enough for food items.

→ Be alert for potential predators, such as outdoor cats or neighborhood hawks.

PROCEDURE

1. Arrange your wild bird food items on the platform, then wait patiently for the first bird to arrive (**Fig. 1**).

2. Use your binoculars to notice details about each visitor, then make a rough sketch in your lab notebook. Pay special attention to beak shape and size. Carefully record which food items the bird tastes and/or eats—they might fly away with it to eat elsewhere (**Fig. 2**).

3. Replenish the food items as they are eaten, and repeat the above process for as many different birds as possible. Did you see any patterns in bird beaks and preferred food items? Which food items were most popular across different bird species? Can you identify the visiting species using a bird guide or online resources?

Fig. 1

Fig. 2

THE
SCIENCE
BEHIND THE FUN

→ Few bird species are
nocturnal like owls; most are
active during daylight hours.
Lab 9 in Unit 2 teaches you
how beak shape influences
food preference, discussing
terms such as seed eaters,
insect eaters, nectar eaters,
and filter feeders. You can apply
that information to species in
the real world. Because birds
are so broadly distributed in
habitats across the globe, it is
easy to interact with them right
around your own home. With
good preparation and careful
observation, you can draw all
sorts of conclusions about
what foods birds prefer. In this
lab, you'll document the feeding
behaviors of visiting birds
and see if beak shape makes
a difference.

CREATIVE ENRICHMENT

→ It's interesting to observe how birds manipulate
food items in different ways. Some just eat an
item all in one bite, others hold items with their feet while
they bite pieces off, some smack items against other
objects to break them apart. Birds are quite intelligent,
and some even use tools to capture their food. Consider
wrapping food items in natural materials to see if birds can
get them open. Try folding a peanut or meal worm into a leaf
and tying it closed with some dried grass. Can visiting birds
open your special package?

BRINGING UP BUTTERFLIES

Explore the life stages of a butterfly before releasing one in your neighborhood.

MATERIALS

→ Caterpillar (from online kit or from the neighborhood)
→ Caterpillar food (from online kit or from the neighborhood)
→ Caterpillar enclosure (from online kit or constructed from a wide-mouthed mason jar and netting or paper towel held in place with a rubber band instead of solid lid)
→ Dry paper towel
→ Dry sticks
→ Handheld magnifying glass (optional)
→ Digital camera or handheld device with camera (smartphone, tablet, etc.)

SAFETY TIPS & HELPFUL HINTS

→ Many online retailers offer butterfly-raising kits, but be sure to order a species that belongs in your region (one that is native to or migrates through your area).

→ You can also go out and find your own caterpillar, but be sure to record and collect leaves from the plant that you found it on. You'll need *lots* of fresh leaves during the larval stage. The best time of year to find your own caterpillar is late spring and summer.

PROCEDURE

1. Line the bottom of the enclosure with a piece of dry paper towel. Caterpillars, just like everything else, produce waste, and the towel makes it easier to clean. Add sticks that are tall enough for a caterpillar to climb and hang off of, but short enough to fit in the enclosure. Sketch your setup in your lab notebook **(Fig. 1)**.

2. Put one or two caterpillars into the enclosure with fresh leaves from the plant on which they were collected or with food from the kit, if you're using one. Be sure to add fresh food and a dry paper towel daily. Make a detailed color sketch of each caterpillar and take photographs. A magnifying glass might be handy for this. In your notebook, also record how long this stage lasts **(Fig. 2)**.

3. Stop feeding the caterpillars once they have all attached themselves to the sticks or netting for pupation. Make a detailed color sketch of each chrysalis and take photographs. Record how long the pupal stage lasts, along with any changes that you notice . Depending on the species, these changes may be obvious or hidden from view **(Fig. 3)**.

4. Make detailed color sketches and take photographs of each butterfly once it hatches and before you release it. If you captured it in the wild, release it back where you found it.

5. Use your sketches and photographs to identify and research the species you raised **(Fig. 4)**.

Fig. 1

Fig. 2

Fig. 3

Fig. 4

THE SCIENCE BEHIND THE FUN

→ Birds are not the only beloved winged visitors making their way through backyards across the globe, there are also butterflies! With more than 18,000 described species, butterflies are found on every continent except Antarctica. They come in different shapes and sizes and exhibit the typical four-stage life cycle of most insects: egg, larva, pupa, and adult (also called *imago*). The eggs adult butterflies lay on plants develop into caterpillars (larval stage). Caterpillars are eating machines, consuming as much food as possible before locating a good site for their next life stage: chrysalis (pupal stage). The changes that occur during the pupal stage are nothing short of incredible. The caterpillar's body *liquifies* and transforms into the elegant but delicate flying insects that have been treasured by human cultures for centuries. In this lab, you'll explore the life stages of a butterfly, from larva to imago, before you send it out into the world.

CREATIVE ENRICHMENT

→ Caterpillars, chrysalises, and butterflies display amazing variation in shape, size, and color. Even butterfly eggs are incredibly diverse! Consider visiting your school library or doing some online research into the world of butterflies, and be sure to share what you discover.

LAB
45

TALKING WITH FROGS

Visit a local water source and try out your frog communication skills.

MATERIALS

→ Flashlight or head lamp
→ Small aquatic net
→ Small plastic or glass container
→ Handheld magnifying glass
→ Digital camera or handheld device with camera (smartphone, tablet, etc.)
→ Phone or handheld device with audio-recording capabilities

SAFETY TIPS & HELPFUL HINTS

→ Frogs generally stick close to water, so try searching for them near ponds, lakes, creeks, and calm river edges. It's best to try this lab on a warm evening in late spring or summer. Scout out your location ahead of time, since part of this lab is done in the dark.

→ Remember that frogs and tadpoles are living creatures, so handle them gently and return them to where you found them. Try not to wear any lotion or insect repellent on your hands, since frogs have permeable skin and can become sick from exposure to such things.

PROCEDURE

1. Head to your frog-watching location right around sunset, settle in a comfortable spot, close your eyes, and listen for frog calls. Make a rough sketch of the landscape in your lab notebook, recording the time of day and an approximate temperature (use your best guess or check the weather). If you hear frog calls, consider recording them **(Fig. 1)**.

2. Put on your headlamp and walk quietly along the water's edge. Watch for any movement from tadpoles swimming in shallow areas near the edge or frogs jumping into the water (you might hear that rather than see it). Once you find an area with tadpole and frog activity, use your net to gently scoop up a tadpole or two. Transfer them carefully into your container and watch them swim around. Sketch a tadpole in your lab notebook. Photograph the tadpole before gently pouring it back into the water source **(Fig. 2)**.

3. Sit quietly without moving at the edge of the activity area and wait for a frog to call. Immediately respond with your own low, raspy, throat sound (do your best to mimic the call you heard). If you get a response, keep calling back and forth at a faster pace each time to see who stops calling first, you or the frogs. How long can you carry on this "combat calling"? Once you get good at it, consider recording yourself talking with the frogs so you can share it.

Fig. 1

Fig. 2

4. Try catching a frog to get a closer look. You'll need to be quick, and be sure to release it after just a few seconds of observation so it doesn't become stressed. Make a sketch or take a photo of any captured frogs to see how different they are from the tadpoles living nearby (have a helper wrangle the camera so it doesn't fall into the water).

CREATIVE ENRICHMENT

→ One way to increase your chances of hosting frogs in your yard is to set up appropriate habitat, such as a frog or toad house. Turn a small flower pot on its side and bury it halfway to create a cozy shelter. Put small rocks and plants around the entrance so it is protected. You can even leave some live meal worms inside the frog house. Check the house each evening to see if anyone has taken up residence!

THE SCIENCE BEHIND THE FUN

→ Butterflies are not the only common backyard visitors that go through significant body changes, there are also frogs! With nearly 7,000 described species living on every continent except Antarctica, frogs make use of many different habitats and exhibit a three-stage life cycle: egg, larva (tadpole), and adult. They vary in size and shape and come in nearly every color you can imagine. Frogs play many important roles in ecosystems as effective predators of pests, such as flies and mosquitoes, indicators for their ecosystem's overall health, and prey items for other animals, from fish and birds to mammals and reptiles—and even other frogs! One particularly interesting characteristic of frogs is their vocal communication. Each species produces its own unique call, which, while primarily intended to attract mates, identifies territories, signals danger, and announces that they are *not* interested in breeding. It's even possible to get frogs to respond to calls made by humans! In this lab, you'll take a closer look at local frogs and test out your call-making skills.

BUILD YOUR OWN FIELD GUIDE

Use your new knowledge to create a field guide for your neighborhood.

MATERIALS

→ Unused notebook or composition book

→ Digital camera or handheld device with camera (smartphone, tablet, etc.)

→ Access to a printer

→ Access to research materials, such as library books and informational websites

→ Colored pencils, crayons, paints, markers

SAFETY TIPS & HELPFUL HINTS

→ Set some limits for yourself for this lab so you don't get overwhelmed. For example, you might want to limit each field guide section to five or fewer species that are present in your area. Consider highlighting three insects, three spiders, three fishes, three amphibians, three birds, three reptiles, and three mammals (you can also include an "Other Invertebrates" category so you can highlight species such as pill bugs and millipedes). You can also limit the area that your field guide focuses on; for example, it can be a field guide to your neighborhood or simply a field guide to your own yard.

PROCEDURE

1. Make a list of species you'd like to include in your field guide, based on what you've recorded in your lab notebook (labs in Unit 1 and Unit 7 will be most useful). Choose the focus of your guide based on your list: will it only highlight one animal group or multiple groups?

2. Dedicate a single page in your unused notebook or composition book to each species. Include a photo or original sketch as well as interesting information about each animal you highlight. What is it called? What does it look like? Where is a good spot to see it? What time of year is it present? What are some of its behaviors? What does it eat **(Fig. 1)**?

3. Be as creative as possible and enjoy sharing your field guide!

Fig. 1

THE
SCIENCE
BEHIND THE FUN

→ Once you have a good feel for the animals living around your home, you can use that knowledge to create a place-based field guide. Explorers have used field guides for over a century, relying on their drawings, photos, and information to learn about the species of plants and animals that live in (or, in the case of some animals, migrate through) a region. Some field guides focus on only one group of animals, such as birds, while others focus on all species, including plants too, found in a defined area. Chances are good that no one has ever created a field guide specifically for your neighborhood though, so through this lab you'll create a one-of-a-kind product!

CREATIVE ENRICHMENT

→ You can also share your neighborhood data through a variety of popular citizen science programs, which harness the help of non-scientists to work on an enormous array of projects. There are active programs all over the world and for all different animal species. Search scistarter.org for projects in your area, and you can share observations about any location in the world using iNaturalist.org. It's quite easy to set up an account and start contributing your observations!

SUPPORTING LOCAL ANIMALS

Wildlife deserves and needs our support. Not only are animals living around your home fascinating to watch and interact with, many of them provide invaluable services, such as seed dispersal, plant pollination, and pest control.

Now more than ever, humans need to create positive change in the world. In the face of habitat alteration, widespread pollution, and global climate change, there has never been a more important time to make responsible decisions that support the animals around us as they struggle to adjust and adapt. The great news is that there are lots of constructive things you can do!

The labs in this unit will help you create a space where local wildlife can thrive. From growing native plants to constructing homes for a variety of local species, this unit gives you the chance to attract and support the animals that call your neighborhood home, doing your part to create a brighter future for wildlife.

GARDEN FOR ANIMALS

Support local animals by designing and building a wildlife-friendly garden.

MATERIALS

→ Native plants
→ Bird bath (base size and shape on your available garden space)
→ Shovel
→ Small pond liner

!

SAFETY TIPS & HELPFUL HINTS

→ The great news is that you can create a wildlife-friendly garden with very little space. Consider setting yours up in your front- or backyard where other plants already exist or on a balcony or patio. You can also transform an existing garden to make it more wildlife-friendly.

→ It's best to design and build your garden in early spring or summer, before it gets too hot for newly installed plants to survive. Consider asking for the supplies needed to create your garden for your birthday or a holiday, or commit to household chores to save up the necessary funds.

PROCEDURE

1. Choose a location for your wildlife garden. Consider access to water for growing plants and filling ponds, access to sun and shade, hardness and composition of the soil if you intend to install plants in the ground, and proximity to heavy foot and car traffic. Take a photo of the space before beginning construction, so you can compare it to the end product **(Fig. 1)**.

2. Choose your plants. Do some research at your school library, a local nature center or museum, or online to identify plants that are native to your area. Staff at local plant nurseries should also be able to help you make informed decisions. Think about each species' mature size, when it blooms, how it smells, which animals it is likely to attract, how it is pollinated, and whether it needs any special soil or nutrients **(Fig. 2)**.

Fig. 1

Fig. 2

THE
SCIENCE
BEHIND THE FUN

→ With habitats across the world being altered by human activities, it's more important than ever to transform available areas to support wildlife. You can design and create valuable wildlife-friendly gardens using your own yard or patio. In return, you get to observe animals up close and benefit from the many services that they provide. The perfect wildlife-friendly garden includes native plant species, food, water, and shelter. The first step is to identify a usable space and research suitable plants. In this lab, you'll start the process of designing and building a wildlife-friendly garden to support local and migrating animal species.

3. Sketch out your ideal garden in your lab notebook, placing each plant species and noting water sources, such as outdoor taps. Will your plants go into the ground or in containers? How often will you need to water your garden? Some plants are always thirsty, but most natives do fine with rainfall once they adjust to their new location. What will you use to get water to the plants (sprinklers, watering can, hose, etc.) **(Fig. 3)**?

4. Work with family, friends, neighbors, and anyone else who is willing to pitch in to turn your design into reality. Depending on the size of the area, you should be able to complete the garden in one day. Be sure to photograph your final product.

Fig. 3

CREATIVE ENRICHMENT

→ It's fascinating (and rewarding) to see how quickly and how often lots of different animals visit your wildlife-friendly garden. Consider recording data about animals' use of the space *before* you create the garden so you can make a true comparison. Watch the space that you intend to use for a few weeks prior to constructing your garden. What kinds of animals use the space and for what purposes? Do you find animal artifacts in the space (fur, feathers, leftover food, bones, scat, etc.)? How quickly do animals begin to use your newly renovated garden? Which added features are most popular with visiting animals?

OH SO SWEET SUGAR WATER

Support your nectar-loving neighbors.

MATERIALS

→ Metal pie tin
→ Rocks of various sizes and shapes
→ Hummingbird feeder
→ Water
→ White cane sugar (do not use brown sugar, raw sugar, honey, or sugar substitutes)
→ Pitcher
→ Binoculars
→ Digital camera or handheld device with camera (smartphone, tablet, etc.)

SAFETY TIPS & HELPFUL HINTS

→ Think carefully about where to put your sugar water feeders. Hang the hummingbird feeder high enough above the ground and away from jumping surfaces so outdoor cats or other predators can't reach it. Also, try to place it where you have a comfortable spot to watch the visitors! Make sure you place the bee feeder in an area with low foot traffic, since some people are scared of bees and others are allergic. Setting it on a rock near flowering plants is a good idea. You might have to relocate your feeders a few times until you find a place where ants don't invade.

→ At least once a week (more often is better, especially in hot weather), scrub out the feeder and rinse it well so no harmful mold develops. Then add fresh sugar water and return the feeder to its spot.

→ Don't bother coloring the water in the hummingbird feeder. Food dyes that are safe for humans might not be safe for birds, and the hummers will be attracted to the feeder without it.

PROCEDURE

1. Make a batch of sugar water by combining one part sugar to four parts clean drinking water and stirring until the sugar dissolves (it can help to start with warm or hot water, but let it cool before putting it out). Start with a small batch (one cup of sugar to four cups of water) and see how long that lasts. You can store leftovers in the refrigerator for a few days, but don't use it if it turns cloudy **(Fig. 1)**.

2. Hummingbird feeder: Fill the hummingbird feeder with sugar water to the very top. It might be best to do this outside in case some spills. Hang it up and watch it often to see who visits! Try to capture a photo of visiting hummingbirds or sketch them in your lab notebook. Did you get more than one species? What time of day were they most active? How quickly did they empty the feeder? How long did they stay at the feeder during each visit? Did any other bird species or animals visit the feeder **(Fig. 2)**?

3. Rinse your rocks with clean water and arrange them to cover the entire bottom of the pie tin (make sure at least some of the rocks are as tall as the rim). Set out the tin where you want it, then carefully pour in sugar water until only the tops

Fig. 1

Fig. 2

→ Once you have a basic wildlife-friendly garden, begin adding other features to attract and support local animals. Consider supplemental food sources, including sugar water for nectar-feeding birds, such as hummingbirds and bees. Bee species across the world are in decline and need everyone's help. Bees are some of the most important pollinators on the planet, supporting the growth and development of many foods we grow, including apples, oranges, cherries, avocados, carrots, pumpkins, and almonds. And wouldn't you be sad to live in a world without broccoli, cauliflower, and Brussels sprouts!? In this lab, you'll work to support your local nectar-loving pollinators.

of some rocks are above the water line. These rocks will serve as little landing platforms for the bees when they visit. Watching from a safe distance with your binoculars, record the bee activity you observe. How long did it take for bees to arrive? How long did the bees typically stay at the feeder? How many bees were there at any one time? Did you observe multiple bee species? Did any other animals visit the feeder **(Fig. 3)**?

Fig. 3

CREATIVE ENRICHMENT

→ Bees don't live on nectar alone—they also need to drink fresh water. Consider making a bee waterer using the same steps you used to create the feeder, only this time use only fresh water. Place it next to the bee feeder so they'll find both important resources. Observe them to figure out which resource is more popular. Do bees visit one and then the other during a single visit? Which do they visit first? What other animals visited the bee waterer?

PINECONE PARADISE

Support visiting birds with custom pinecone feeding stations.

SAFETY TIPS & HELPFUL HINTS

→ Hang your pinecone feeders out of reach of potential predators in a quiet area with shade (so your peanut butter doesn't melt in the sun).

→ If you don't have pinecones in your area, order them online or buy them at a local craft store (just make sure they have not been painted or sprayed with chemicals).

PROCEDURE

1. Clean your pinecones by shaking them and removing any dirt or debris. You may want to use gloves, since pine sap can be very sticky and difficult to remove from skin and clothing.

2. Tie string around the center of each pinecone so you end up with a 2-to-3-foot (0.6 to 0.9 m) hanging length on each. Use strong knots—the pinecones will get quite heavy as you add food **(Fig. 1)**.

3. Layer on peanut butter everywhere on each pinecone with a clean, dull knife. Working it far into the grooves and rows of the scales **(Fig. 2)**.

4. Spread fresh bird seed onto a plate or the bottom of a pie tin. Roll the peanut butter-covered pinecones into the seed to cover all sides. Add some extra delicious food items (large nuts, dried fruit, etc.) to each pinecone **(Fig. 3)**.

5. Hang your pinecone feeders and have a seat nearby (these feeders are so tasty that birds will likely let you get pretty close). Record how long it takes to attract the first visitors. Use your binoculars to get a closer look and take time to sketch the birds in your lab notebook. What was the most common species that visited? Did other animals visit the pinecone feeders?

MATERIALS

→ Natural pinecones (no paint, glitter, scents, etc.)
→ String
→ Scissors
→ Peanut butter
→ Variety of bird-friendly foods (nuts, sunflower seeds, dried fruit, etc.)
→ Bird seed
→ Plate or pie tin
→ Binoculars

Fig. 1

Fig. 2

Fig. 3

THE
SCIENCE
BEHIND THE FUN

→ Whether you're interested in documenting the presence of a certain species or just curious about the variety of birds that live in your area, setting out and maintaining feeders can be a lot of fun. But there is another important reason to construct and set out bird feeders: birds perform some of the longest and most tiring migrations in the animal world. For example, the Arctic tern travels more than 40,000 miles per year between Greenland and Antarctica. All that travel can be pretty exhausting, and migrating birds must make use of whatever food sources are available as they make stops along the journey. Avid birders around the world actively support species by setting out food along migratory paths. Feeders can be made from many different materials and contain a variety of food items, because, as you've discovered from previous labs, birds eat all kinds of things (nectar, fishes, insects, nuts, shrimps, seeds, fruits, etc.). In this lab, you'll use a simple natural structure (pinecone) to create attractive food stations and support birds that are passing through, as well as local species.

CREATIVE ENRICHMENT

→ Peanut butter isn't the only base material out there. Consider using almond butter, coconut oil, suet, lard, and vegetable shortening. Which material is most popular with visiting birds? Also consider hanging these feeders at different times throughout the year to see if the species vary across seasons.

BLISSFUL BIRDHOUSE

Support local birds with a custom-built birdhouse.

birdhouse rather than a decorative one. Entrance hole sizes, for example, will vary, but it should never be big enough to let in predators.

PROCEDURE

1. Build your birdhouse according to the kit instructions, and ask for help if you need it. Every birdhouse should have small drainage holes in the floor to keep moisture from pooling and small ventilation holes just under the roofline. Consider removing the outside perch, if there is one, as it's unnecessary and might help unwanted animals, such as squirrels, get inside. If possible, try to modify your birdhouse so you can easily open it for regular cleanings (hinges and a latch for the roof or floor work well) **(Fig. 1)**.

2. Attach natural materials to your birdhouse with non-toxic, waterproof glue. Many kits include paint for decorating the birdhouse, but natural colors and natural items actually work better to attract birds. Just don't use anything too heavy. Sketch your final product in your lab notebook **(Fig. 2)**.

MATERIALS

→ Birdhouse kit
→ Common tools (screwdriver, power drill, hammer, etc.)
→ Natural materials (tree bark, moss, dried leaves, acorns, pine needles, etc.)
→ Non-toxic, waterproof glue
→ Mounting rope or chain

SAFETY TIPS
& HELPFUL HINTS

→ If someone at your house has woodworking tools and experience, consider designing and constructing your own simple birdhouse (look for free plans online). For most people, it's easiest to construct a birdhouse from a kit purchased at a craft store or online. Be sure to select a functional

Fig. 1

Fig. 2

3. Hang your birdhouse from a tree branch, so it hangs at least 1 foot (0.3 m) below the branch, in a quiet, shady spot away from feeders and other bird hot spots. Make sure that it is out of reach of squirrels and other climbing animals, as well as any predators coming from the ground. Check in on your birdhouse now and then to see if any birds are using it—it might take several days or even weeks to attract a resident **(Fig. 3)**.

Fig. 3

→ Supporting local birds is about more than just being a good neighbor. Birds help pollinate plants and distribute their seeds, and their diets make them useful for controlling weeds and pest populations. Setting out feeders is a great way to attract birds, but a birdhouse can convince them to linger. Depending on the time of year, a birdhouse may provide a nursery for chicks, a safe overnight roost, a shelter from bad weather, or an escape from predators. A small to medium-sized birdhouse will likely attract wrens, chickadees, sparrows, and bluebirds. In this lab, you'll provide local birds with safe haven by constructing and decorating a simple birdhouse.

CREATIVE ENRICHMENT

→ Even though you have supplied a birdhouse, bird pairs will likely still build a simple nest inside and therefore need access to appropriate materials. Nesting materials are important for cushioning eggs and chicks and for providing warmth. Consider adding some nesting material, such as dried grasses or leaves, feathers, twigs, pine needles, yarn and other natural fibers (cut into short pieces so birds don't become tangled), animal fur, and moss, to the inside of the birdhouse. Also leave some of these materials on the ground below the birdhouse and around your yard so bird pairs will have plenty from which to choose.

OUTSTANDING OWL BOX

Support local barn owls with a specially designed owl box.

MATERIALS

→ Barn owl nest box kit
→ Common tools (screwdriver, power drill, hammer, etc.)
→ Wood stain
→ Paintbrush
→ Mounting pole or large tree trunk and hardware

SAFETY TIPS & HELPFUL HINTS

→ If someone at your house has woodworking tools and experience, consider designing and constructing your own barn owl box (look for free plans online). For most people, it's easiest to construct an owl box from a kit purchased online. For barn owls, it's best to work with a design that results in a rectangular box that is at least 24 inches (61 cm) long, 18 inches (45.7 cm) high, and 18 inches (45.7 cm) wide, with an entrance hole diameter of 5 inches (12.7 cm).

Fig. 1

Fig. 2

→ Having owls around your home and in your neighborhood helps maintain productive gardens as owls do a great job eating rodents, such as rats, mice, and squirrels, that often cause damage to crops and other property. One of the most widely distributed groups of bird species on the planet is the barn owls, which are found on every continent except Antarctica. They make excellent neighborhood rodent hunters, easily adapting to rural, suburban, and urban settings. Owls require a nest box with specific design parameters. In this lab, you'll design and build a barn owl box to support these majestic (and helpful) birds of prey.

PROCEDURE

1. Build your owl box according to the kit instructions, and ask for help if you need it. Make sure to include small drainage holes on the bottom and small ventilation holes just under the roofline. If possible, try to modify your owl box so you can easily open it for regular cleanings (hinges and a latch for the floor or roof work well) **(Fig. 1)**.

2. Consider using wood stain to seal and weatherproof your owl box. Sketch your final product in your lab notebook **(Fig. 2)**.

3. Mount your owl box to the trunk of a large tree in a quiet, shady spot or atop a sturdy metal mounting pole out in your yard. Regularly check on the box to see if any owls are using it, especially in the evening around sunset. How long did it take for an owl to use the box? What time of year was the box most popular **(Fig. 3)**?

Fig. 3

CREATIVE ENRICHMENT

→ Establishing an owl box in your yard can also provide endless numbers of owl pellets, which can inspire many exciting research questions. Consider designing an original study or science fair project around owl pellets (the regurgitated, indigestible parts of an owl's food, such as bones and hair). What is the most common prey item for owls in your yard? How many animals do owls eat in a week? A month? Lots of interesting questions!

BENEFICIAL BAT HOUSE

Support local bat species with a bark-covered bat house.

MATERIALS

→ Bat house kit
→ Common tools (screwdriver, power drill, hammer, etc.)
→ Wood stain
→ Paintbrush
→ Non-toxic, waterproof glue
→ Tree bark
→ Mounting hardware

SAFETY TIPS & HELPFUL HINTS

→ If someone at your house has woodworking tools and experience, consider designing and constructing your own bat house (look for free plans online). For most people, it's easiest to construct a bat house from a kit purchased online.

→ Be sure not to handle bats or their droppings (also called guano), as some species can carry rabies. Consider placing a disposable ground covering, such as a plastic trash bag or butcher paper, below the bat house to collect guano for disposal.

PROCEDURE

1. Build your bat house according to the kit instructions, and ask for help if you need it. Before the house is complete, use a screwdriver or chisel to rough up the wood that will become the inside of the house and make it feel more like tree bark **(Fig. 1)**.

2. Consider using wood stain to seal and waterproof your bat house. You may also glue pieces of tree bark to the outside of the bat house to make it look more natural and to give visiting bats something to grip. Sketch your final product in your lab notebook **(Fig. 2)**.

3. Mount your bat house to the side of your house or other outdoor structure in a quiet, sunny spot with good flight access from below (bats will fly down in front of the house and then up into the cavity). Regularly check on the house to see if any bats are using it, especially in the evening around sunset. How long did it take for a bat to use the house? What time of year was the house most popular **(Fig. 3)**?

Fig. 1

Fig. 2

Fig. 3

→ There are more than 1,200 species of bats found nearly worldwide. A wonderful way to support your local bat species is by putting up a bat house. Bats typically find a nice, cozy space under the bark of a tree to give birth and raise babies, but as humans continue to alter global habitats, there are fewer and fewer suitable trees in many areas of the world. Building a bat house is a great way to encourage bats in your area to take shelter in your yard as they raise their young and control pesky mosquito populations. In this lab, you'll build and mount a bat house to support these incredible flying mammals.

CREATIVE ENRICHMENT

→ Bats are marvelously diverse, representing roughly 20 percent of all mammal species on the planet. Chances are good that several different species live around your house, so take some time to visit your school library or do some online research to become familiar with the bats in your area. Use your binoculars to watch the bats that fly in and out of your bat house. Can you identify the species? Pay special attention to their size and color when trying to remember important body features that will help with identification. And remember to share what you learn!

TAKING CONSERVATION ACTION

Exploring the infinite wonders of the animal kingdom should inspire you to assess your own impacts on the environment. When you appreciate the unique and fragile beauty of the animals around you, it's easy to advocate on their behalf. Conservation begins with individual actions, so here are some simple things you can do to create real change for wildlife.

HELP CONSERVE WATER

One of the world's most important (and limited) natural resources is fresh water. Every living organism on this planet needs it in order to survive, so we must do our best to use it wisely. You can conserve water at home, at school, and at play—anywhere you go. Think through your own daily water use and identify areas where you can use less. Taking shorter showers, using strategic watering regimes for outdoor plants and lawns, and turning off the faucet while you brush your teeth are simple ways to reduce your impact on our global water supply and help ensure that your favorite animals have what they need, too.

JUST SAY NO TO PLASTICS

Nowadays plastics pollution affects nearly every species on the planet. Whether it is accidentally ingested by animals, such as sea turtles and whales, or accumulates in the bodies of animals up through the ocean food chain, plastics—especially single-use plastics—have long-term negative impacts on animals and their habitats, including humans! The next time you're out to eat with your family, simply say no to plastic straws when they're offered. When you go shopping, remember your reusable bags. And go one step further by examining single-use plastics in all areas of your life. How can you reduce those uses, too? The animals would thank you if they could!

KEEP YOUR CAT INDOORS

You may love cats and, perhaps, even have more than one, but it is important to know that they can cause great harm if let outdoors; in fact, cats are responsible for killing *billions* of birds and small mammals across the globe every year. In addition to the possible harm they can cause, cats risk being hit by cars, picking up parasites or diseases, fighting with other cats, and becoming prey, among other things. Keeping your cat(s) indoors is safer and healthier for them, and it helps protect local wildlife. Think about it!

EAT LESS MEAT

As delicious as cheeseburgers are, the production of patties (and other types of meat) takes a heavy toll on the environment. Many biodiverse habitats around the world are destroyed to create space for grazing cattle, a process that also contributes to climate change and other forms of environmental pollution. But if you're not ready to go completely vegetarian, even just reducing your meat consumption can help a great deal. Try going meatless just one day per week or only eating meat with one meal each day. This small action will have a big impact on animals and their habitats.

HELP TO COMBAT CLIMATE CHANGE

Climate change is affecting species all across the planet, including humans. We are all experiencing warmer temperatures, increased periods of drought, and more frequent and severe weather events such as storms, floods, and wildfires. The vast majority of world scientists agree that humans are causing these changes, mostly by burning fossil fuels to power our communities. Since humans are the cause, we can also be the solution, by changing our behavior. You can combat climate change by conserving electricity (turn off lights when you leave a room, unplug electronics when they're not in use, and adjust your thermostat to a more narrow temperature range) and also by walking and biking to places or by taking a bus or train.

REDUCE YOUR CHEMICAL USE

Now that you have observed firsthand how you share your local ecosystems with many other fascinating creatures, it's important to make responsible decisions about what you put into the environment. Next time you or a family member considers the use of synthetic pesticides or herbicides to ward off unwanted species, try researching nature-based alternatives instead. For example, non-toxic soaps, oils, and other ingredients work wonders for discouraging pest species without harming the environment or your family! Better yet, try integrated pest management, which involves using beneficial animals and plants to deter and control the undesirable ones.

DON'T DISTURB THE PEACE

Have you ever been peacefully enjoying yourself while watching gorillas at the zoo when a super obnoxious person walks up and yells at them to get their attention? The actions of humans have major impacts on wildlife behavior, so be sure to stay respectful and peaceful when it comes to sharing space with animals, no matter where you find them.

ACKNOWLEDGMENTS

I offer my genuine thanks to Jonathan Simcosky, David Martinell, Meredith Quinn, Andrea Zander, and the whole Quarry Books team for collaborating with me on this wonderful project! Researching and writing about animals made me remember all over again why I love wildlife and why I will work for the rest of my life to conserve it.

I am also extremely grateful to San Diego Zoo Wildlife Alliance, most especially Allison Alberts, Ollie Rider, and the entire Community Engagement team (aka League of Extraordinary Women) for supporting me in my professional and personal pursuits. How lucky I am to work for an organization with such an inspiring mission and such a bold vision! Thanks also to the San Diego Natural History Museum, Cardiff Elementary School, and the National Science Foundation for igniting my love of science teaching.

A special thanks to Marshal Hedin for inspiring me to be a champion for invertebrates (the little things that run the world), and to Barry and Anne Munitz for opening my eyes to the world and for their incredible support of our team's community work here and abroad.

I would also like to thank my parents for so frequently taking us out into nature as kids (from Redfish Lake to Silver Falls) and my sister, Jenna, for always disappearing on long hikes … you taught me that there is always more to discover (you are still teaching me that today!). Also a special thanks to my sweet Poppy for his hours of dedicated, expert proofreading! Thank you to all the wonderful children who shared their smiles for this book (and, of course, their awesome, patient parents)! Lastly, love and thanks to my husband, Brad (an amazing photographer among many other things), and my daughters, Wren and Phoebe, for reminding me what's most important in life, and for always being up for adventures in nature. Here's to many more years of exploring!

ABOUT THE AUTHOR

Maggie Reinbold is Director of Community Engagement at San Diego Zoo Wildlife Alliance, heading up a dynamic team dedicated to designing and implementing programs that connect communities to conservation for the benefit of wildlife and habitats. She is also an adjunct faculty member in the Department of Biology at Miami University, where she teaches the Earth Expedition to the Big Island of Hawaii. Maggie earned her bachelor's and master's degrees in biology at San Diego State University, with a focus on the population genetics of desert aquatic insects across the Baja California Peninsula.

She has taught science in a number of formal and informal settings, including the San Diego Zoo Institute for Conservation Research, San Diego Natural History Museum, Cardiff Elementary School, and San Diego State University. As a National Science Foundation science fellow, she co-taught hands-on science with classroom teachers across San Diego County and also spent several seasons in Arctic Alaska, bringing hands-on science education to unique and underserved communities on the North Slope. She lives with her husband and daughters in Poway, California.

ABOUT THE PHOTOGRAPHER

Bradford Hollingsworth is a professional biologist and museum curator at the San Diego Natural History Museum. His lifelong passion for photography started at an early age and stemmed from a childhood speech disorder that redirected his primary form of communication from verbal to visual. Photography became a natural outlet, which soon merged with his love of nature. Brad specializes in animal photography, capturing people exploring nature, and imaging natural history specimens from the museum's vast collections. His photography has appeared in nature guides, marketing materials, websites, books, and museum exhibitions. Brad earned his bachelor's and master's degrees from San Diego State University and his doctorate from Loma Linda University.

The author and photographer are a married couple with two children and enjoy exploration of the natural world as a family. Many of the book's activities were done as science projects at home and as part of the family's expeditions to the world's forests, jungles, oceans, and deserts.

Alexandra

Ava

Bailey

Bennett

Brennan

Caleb

Clara

Elsie

Faith

Halle

Hana

Jacqueline

Karim

Lauren

Mariam

Maris

Nicholas

Phoebe

River

Samantha

Sarah

Sophia

Troy

Tyler

Vincent

Wren

Lou Lou

INDEX